Highway-Runoff Quality, and Treatment Efficiencies of a Hydrodynamic-Settling Device and a Stormwater-Filtration Device in Milwaukee, Wisconsin

By Judy A. Horwatich, Roger T. Bannerman, and Robert Pearson

Prepared in cooperation with the
Wisconsin Department of Transportation
and the Wisconsin Department of Natural Resources

Scientific Investigations Report 2010–5160

U.S. Department of the Interior
U.S. Geological Survey

U.S. Department of the Interior
KEN SALAZAR, Secretary

U.S. Geological Survey
Marcia K. McNutt, Director

U.S. Geological Survey, Reston, Virginia: 2011

For more information on the USGS—the Federal source for science about the Earth, its natural and living resources, natural hazards, and the environment, visit http://www.usgs.gov or call 1-888-ASK-USGS

For an overview of USGS information products, including maps, imagery, and publications, visit http://www.usgs.gov/pubprod

To order this and other USGS information products, visit http://store.usgs.gov

Contents

Figures

Tables

Conversion factors, abbreviations, and acronyms

Multiply	By	To obtain
Length		
inch (in.)	2.54	centimeter (cm)
foot (ft)	0.3048	meter (m)
Area		
acre	4,047	square meter (m^2)
square foot (ft^2)	0.09290	square meter (m^2)
Volume		
cubic foot (ft^3)	0.02832	cubic meter (m^3)
Flow rate		
cubic foot per second (ft^3/s)	0.02832	cubic meter per second (m^3/s)
inch per hour (in/h)	0.0254	meter per hour (m/h)
Mass		
pound, avoirdupois (lb)	0.4536	kilogram (kg)

Concentrations of chemical constituents in water are given either in milligrams per liter (mg/L) or micrograms per liter (µg/L).

Particle sizes of sediment are given in micrometers (µm). A micrometer is one-thousandth of a millimeter.

Abbreviations used in this report

Ca	Calcium
Cl	Chloride
COD	Chemical oxygen demand
DCu	Dissolved copper
DP	Dissolved phosphorus
DZn	Dissolved zinc
EMC	Event-mean concentration
ETV	Environmental Technology Verification
GMIA	General Mitchell International Airport
HSD	Hydrodynamic-settling device
I–794	Interstate 794
LOQ	Limit of quantification
Mg	Magnesium
MOU	Memorandum of Understanding
NOAA	National Oceanic and Atmospheric Administration
NSF	National Sanitation Foundation (International)
NURP	Nationwide Urban Runoff Program
PAH	Polycyclic aromatic hydrocarbon
PSD	Particle-size distribution

QA/QC	Quality-assurance/quality-control
RPD	Relative percent difference
SFD	Stormwater-filtration device
SOL	Summation of loads
SRS	Standard reference sample
SS	Suspended sediment
TCu	Total copper
TDS	Total dissolved solids
TP	Total phosphorus
TSS	Total suspended solids
TZn	Total zinc
USEPA	U.S. Environmental Protection Agency
USGS	U.S. Geological Survey
VA	Visual accumulation
WDNR	Wisconsin Department of Natural Resources
WisDOT	Wisconsin Department of Transportation
WSLOH	Wisconsin State Laboratory of Hygiene

Highway-Runoff Quality, and Treatment Efficiencies of a Hydrodynamic-Settling Device and a Stormwater-Filtration Device in Milwaukee, Wisconsin

By Judy A. Horwatich[1], Roger T. Bannerman[2], and Robert Pearson[3]

Abstract

The treatment efficiencies of two prefabricated storm-water-treatment devices were tested at a freeway site in a high-density urban part of Milwaukee, Wisconsin. One treatment device is categorized as a hydrodynamic-settling device (HSD), which removes pollutants by sedimentation and flotation. The other treatment device is categorized as a stormwater-filtration device (SFD), which removes pollutants by filtration and sedimentation. During runoff events, flow measurements were recorded and water-quality samples were collected at the inlet and outlet of each device.

Efficiency-ratio and summation-of-load (SOL) calculations were used to estimate the treatment efficiency of each device. Event-mean concentrations and loads that were decreased by passing through the HSD include total suspended solids (TSS), suspended sediment (SS), total phosphorus (TP), total copper (TCu), and total zinc (TZn). The efficiency ratios for these constituents were 42, 57, 17, 33, and 23 percent, respectively. The SOL removal rates for these constituents were 25, 49, 10, 27, and 16 percent, respectively. Event-mean concentrations and loads that increased by passing through the HSD include chloride (Cl), total dissolved solids (TDS), and dissolved zinc (DZn). The efficiency ratios for these constituents were –347, –177, and 20 percent, respectively. Four constituents—dissolved phosphorus (DP), chemical oxygen demand (COD), total polycyclic aromatic hydrocarbon (PAH), and dissolved copper (DCu)—are not included in the list of computed efficiency ratio and SOL because the variability between sampled inlet and outlet pairs were not significantly different.

Event-mean concentrations and loads that decreased by passing through the SFD include TSS, SS, TP, DCu, TCu, DZn, TZn, and COD. The efficiency ratios for these constituents were 59, 90, 40, 21, 66, 23, 66, and 18, respectively. The SOLs for these constituents were 50, 89, 37, 19, 60, 20, 65, and 21, respectively. Two constituents—DP and PAH— are not included in the lists of computed efficiency ratio and SOL because the variability between sampled inlet and outlet pairs were not significantly different. Similar to the HSD, the average efficiency ratios and SOLs for TDS and Cl were negative.

Flow rates, high concentrations of SS, and particle-size distributions (PSD) can affect the treatment efficacies of the two devices. Flow rates equal to or greater than the design flow rate of the HSD had minimal or negative removal efficiencies for TSS and SS loads. Similar TSS removal efficiencies were observed at the SFD, but SS was consistently removed throughout the flow regime. Removal efficiencies were high for both devices when concentrations of SS and TSS approached 200 mg/L. A small number of runoff events were analyzed for PSD; the average sand content at the HSD was 33 percent and at the SFD was 71 percent. The 71-percent sand content may reflect the 90-percent removal efficiency of SS at the SFD. Particles retained at the bottom of both devices were largely sand-size or greater.

Introduction

In Wisconsin, State and Federal regulations apply to the quality of stormwater runoff from the State highway system. The Wisconsin Department of Transportation (WisDOT) finalized a Memorandum of Understanding (MOU) in 1994 with the Wisconsin Department of Natural Resources (WDNR) for the control of stormwater-runoff flows from the highway system (Wisconsin Department of Transportation, 2002). The MOU covers State-owned and -operated systems in Milwaukee and Madison and includes a phased approach to examine stormwater-control opportunities at many other municipal areas. In addition, the U.S. Environmental Protection Agency (2000) Phase II stormwater regulations focus on the quality of water flow from storm sewers.

The MOU requires at least an 80-percent reduction in total suspended solids (TSS) for transportation facilities first constructed on or after January 2003 and a maximum extent possible for reconstructive highway projects (Wisconsin

---

[1] U.S. Geological Survey.

[2] Wisconsin Department of Natural Resources.

[3] Wisconsin Department of Transportation.

Department of Transportation, 2002). The cost of land in high-density urban areas can be prohibitive for implementation of traditional stormwater systems such as wet detention basins. Alternatives include prefabricated-treatment devices that are more compact and can be installed underground. The pollutant-removal efficacies of these stormwater-treatment devices have not been tested previously for direct field applications in Wisconsin. The study described in this report evaluated the effectiveness and practical applications for two of the many prefabricated-treatment devices designed to improve the quality of stormwater runoff.

This study builds on a long history of U.S. Geological Survey (USGS) urban water-quality investigations in Wisconsin. In 1978, the U.S. Environmental Protection Agency (USEPA) established the Nationwide Urban Runoff Program (NURP) to assess the water-quality characteristics of urban runoff (Bannerman and others, 1983). When the City of Milwaukee, Wis., was chosen by the USEPA as a NURP site, a partnership between the WDNR and the USGS was developed to evaluate urban runoff in Milwaukee. Since the NURP study, the USGS and the WDNR have continued their partnership and have completed more than 15 studies in at least 6 cities to assist the State of Wisconsin in characterization of urban stormwater runoff (appendix 1). Results from this study provide additional information to meet the partnership goals of understanding urban runoff.

In 1999, the USEPA established the Environmental Technology Verification (ETV) program, setting a national focus on validating the performance of technologies that includes verifying manufacturers' claims for efficiency of prefabricated-treatment devices. The USEPA, with the National Sanitation Foundation International (NSF International) as its verification partner, is charged with the following tasks: (1) create a national protocol to test wet-weather flow technologies, (2) contract independent groups to evaluate the effectiveness of the stormwater-treatment devices of interest, (3) review and implement the verification-testing plans, and (4) make study results available to the general public (U.S. Environmental Protection Agency, 2002). Municipalities and other interested parties will then have access to all ETV program results to assist them in making informed decisions on the choice of stormwater-treatment devices for their stormwater-management programs. Results from this study have been forwarded to ETV personnel for their final verification reporting (U.S. Environmental Protection Agency, 2004; 2005b).

As part of their efforts to improve the quality of highway runoff, the WisDOT has worked in cooperation with the USGS, WDNR, City of Milwaukee, the Milwaukee Third Ward, Milwaukee County, and with the USEPA ETV program to verify the treatment efficiencies of two prefabricated stormwater-treatment devices. The cooperators shared in either the cost of installing the devices or the cost of monitoring them. In December 2001, two devices were installed in a Milwaukee County parking lot beneath an elevated freeway—Interstate 794 (I–794)—in Milwaukee (fig. 1), next to the Milwaukee River. These monitoring stations have been referred to as the "Milwaukee Riverwalk Sites." Both devices were connected to pipes draining a section of the freeway; the devices had been installed in Wisconsin previously but had never been evaluated for their effectiveness.

These devices are 2 of 10 such stormwater-treatment devices that the WDNR and USGS have examined to evaluate water-quality effects. A third study was done in cooperation with the ETV program (Bachhuber and others, 2001; Horwatich and others, 2004).

Purpose and Scope

The primary objective of this report is to describe the effectiveness of two prefabricated-treatment devices in removing a suite of inorganic and organic water-quality constituents from stormwater runoff. This report also describes methods and techniques used to determine the effectiveness of these devices. Detailed data describing water quality, flow, constituent loads, and removal efficiencies are presented for inlet and outlet samples collected between June 2002 and October 2004.

Another objective of this report is to add to the understanding of stormwater-runoff quality and quantity in an urban environment. The USGS and the WDNR have cooperated in many projects that help characterize quality and quantity of urban runoff. The results of these studies have assisted State and Federal agencies in making informed stormwater-management decisions (appendix 1).

Site Description

In December 2001, two devices were installed in a municipal parking lot beneath an elevated span of I–794 (fig. 1). The parking lot is west of Water Street, between Clybourn Street and St. Paul Avenue, in downtown Milwaukee. Runoff flowed from the devices directly to the Milwaukee River upstream from the mouth to Milwaukee Harbor, which flows into Lake Michigan.

The climate of Milwaukee and Wisconsin in general, is typically continental with some modification by Lake Michigan. Milwaukee experiences cold, snowy winters and warm to hot summers. Average annual precipitation is approximately 32 in., and average annual snowfall is 47.5 in. (National Oceanic and Atmospheric Administration, 2007a,b).

During the winter, snow and ice is removed from freeways by the use of road salt. The freeway is cleaned by a conventional (mechanical) sweeper once per month and by special assignment (such as when a truck spills debris on the freeway). The Milwaukee metropolitan area has the largest density of traffic in the State of Wisconsin. Vehicle exhaust and other industrialized factors contribute to the USEPA designating Milwaukee County as a non-attainment area for high ozone levels during the summer. When the ozone exceeds 85 parts per billion alternate forms of transportations are recommended.

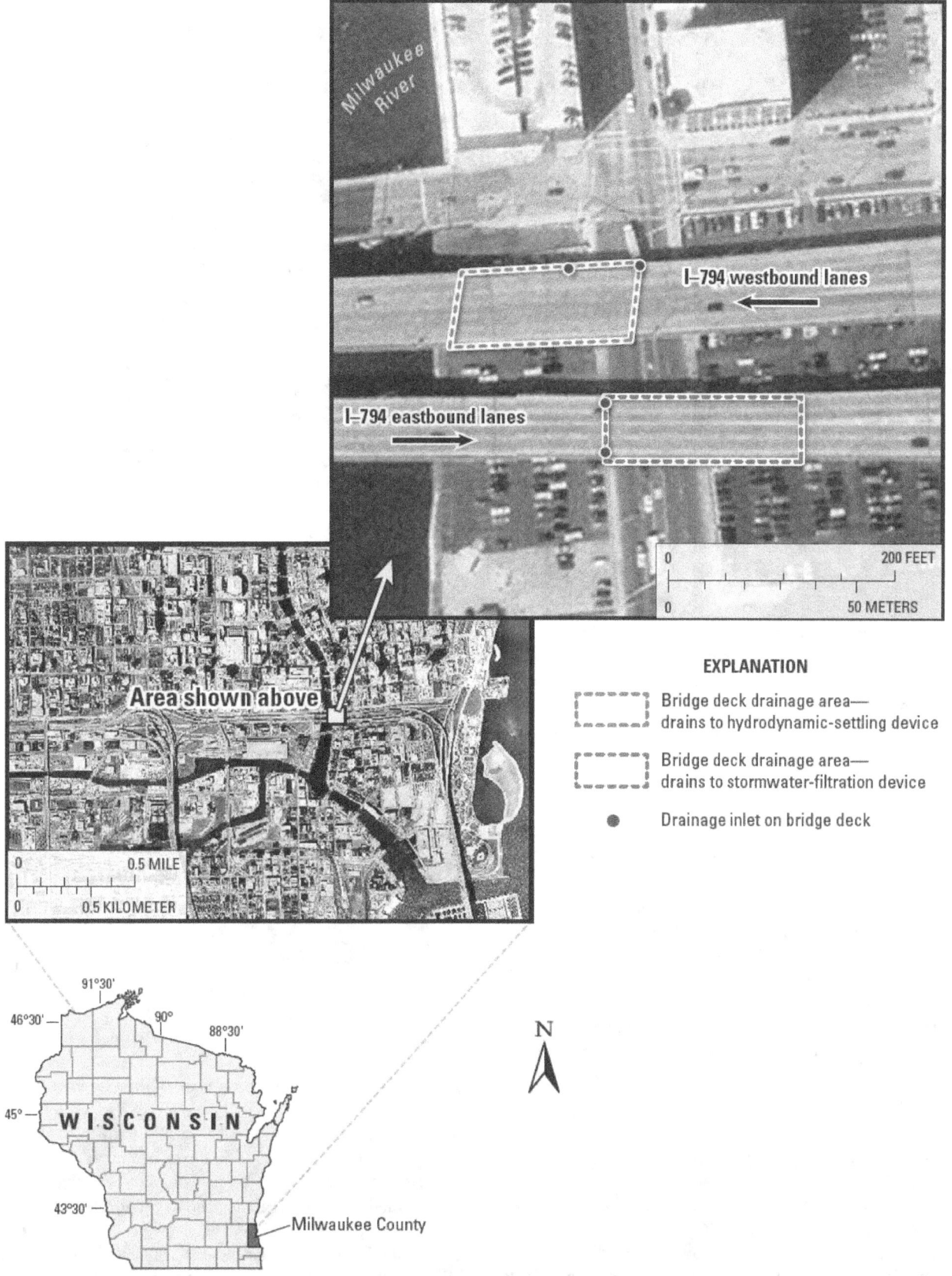

Figure 1. Location of monitored sites for the hydrodynamic-settling and stormwater-filtration devices in the City of Milwaukee, Wisconsin. Photographs by the Wisconsin Department of Transportation.

Figure 2.   Piping system from the westbound I–794 freeway to the hydrodynamic-settling device.

The eastbound and westbound decks of I–794 were originally constructed in 1967 and were last overlaid in 1993 with a bituminous surface. The condition of the elevated freeway was rated as "poor" during the time of the study, and reconstruction of the freeway was planned for 2007. The average daily-traffic count during the study period was 47,000 vehicles.

Hydrodynamic-Settling Device

The hydrodynamic-settling device (HSD) treats a 0.25-acre deck section of the westbound I–794 freeway, encompassing five lanes and an outside shoulder (fig. 1). The drainage surface on the westbound freeway slopes gradually eastward (0.5-percent slope) and dips slightly to the north. Runoff flows across the lanes toward the outside edge of the deck into two storm-drain inlets on the north side of the freeway deck. Two 6-in.-diameter downspouts then connect into 8-in. piping connected to the device. Segments of the 8-in. pipe are on a slope of 5.6 percent and are approximately 15 ft above the parking lot (fig. 2).

Stormwater-Filtration Device

The stormwater-filtration device (SFD) treats a 0.19-acre deck section of the eastbound I–794 freeway, encompassing four driving lanes and an outside shoulder. The drainage surface slopes gradually westward (1.7 percent) and dips slightly to the south. The two storm drains are across from each other on opposite sides of the deck (fig. 1). Runoff entering the

inlets drops into 6-in.-diameter downspouts that connect to an 8-in. pipe. The downspouts are on a slope of 5.6 percent and are approximately 15 ft above the parking lot. The 8-in. connection pipe drops 6 ft to the ground surface and then another 4 ft below the ground, which drains into a 9-ft length of lateral pipe connected to the device.

Description of the Hydrodynamic-Settling and Stormwater-Filtration Devices

These devices use different processes to treat stormwater runoff. The HSD removes pollutants by sedimentation and flotation. This study focused on the settling processes but did not attempt to measure floating material, such as large quantities of trash and oil. The HSD has a circular swirl chamber that causes a rotating flow field to remove the heavier particles; a floatable baffle wall to entrap surface oil, grease, and floating material; and low-flow and high-flow weirs for discharging flows. The SFD removes pollutants by filtration and sedimentation. Filtration is considered the primary method of treatment; a filter media is used to retain the pollutants by sorption. Sedimentation of larger particles occurs in a pretreatment chamber and on the bottom of the cartridge-filter bay. When flows exceed the peak design rate, the SFD is designed to allow untreated water to bypass the cartridge filtration.

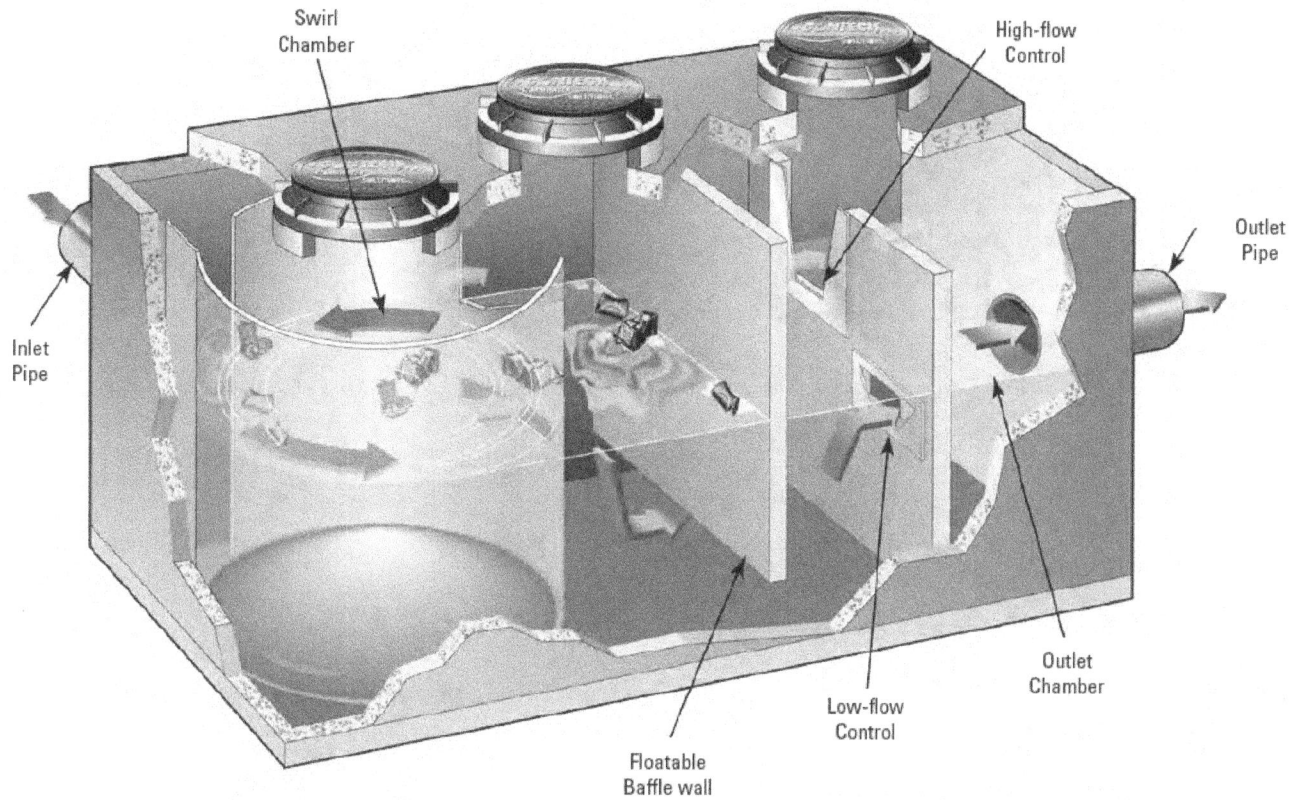

Figure 3. Schematic diagram of the hydrodynamic-settling device (J.H. Lenhart, written comm., 2008). Image courtesy of Contech Construction Products Inc., used with permission.

Hydrodynamic-Settling Device

The HSD station was commonly referred to as "Riverwalk North" because it was at the north end of the municipal parking lot. Station identification numbers and names for the HSD inlet and outlet monitoring sites are 430208087543201, Milwaukee Riverwalk North Device inlet at Milwaukee; and 430209087543200, Milwaukee Riverwalk North Device outlet at Milwaukee.

The device was housed in a 6-in. thick, 10-ft long, 3-ft wide, and 8-ft deep concrete structure (fig. 3). The 10-ft length of pipe connected to the HSD was considered part of the device because the device created backwater in the pipe allowing sediment to drop out into the pipe. The runoff flows from the inlet pipe into a 3ft-diameter swirl chamber, which is the principal settling unit. Past the swirl chamber, a floatable baffle wall extends from the top of the device to 6 in. above the floor to trap oil and floating material. The flow-control wall has two weirs; a low-flow weir set at an elevation of 3 ft above the floor and the high-flow weir set at an elevation of 4.9 ft above the floor (U.S. Environmental Protection Agency, 2005b). The weirs are designed to create backwater to increase efficiency of the device. All flow exits through an 8-in. pipe.

Design peak capacity is approximately 0.27 ft³/s per square foot of swirl-chamber area. This device was designed to treat flows with a peak flow rate of 1.6 ft³/s. It was not designed with a bypass, so flows exceeding 1.6 ft³/s go over the high-flow weir wall, decreasing settling time through the device.

Stormwater-Filtration Device

The SFD station was commonly referred to as "Riverwalk South" because it was at the south end of the municipal parking lot. Station identification numbers and names for the inlet and outlet monitoring sites are 430207087543200, Milwaukee Riverwalk South device inlet at Milwaukee; and 430208087543200, Milwaukee Riverwalk South outlet at Milwaukee.

The SFD was housed in a 6-in. thick, 12-ft long, 6-ft wide, and 5.5-ft deep concrete structure (fig. 4). Inlet flow enters a 2-ft-wide and 1.67-ft-deep inlet bay where the larger particles are intended to drop out. Runoff then flows through a flow spreader that disperses water evenly into a 7.4-ft-long filtration bay. The nine filter cartridges for this study were designed to remove sediments, metals, organics, phosphorous, oils, and greases.

Each cartridge was 1.5-ft high and was filled with ZPG media, a mixture of zeolite, perlite, and granular activated carbon (U.S. Environmental Protection Agency, 2004). Flow

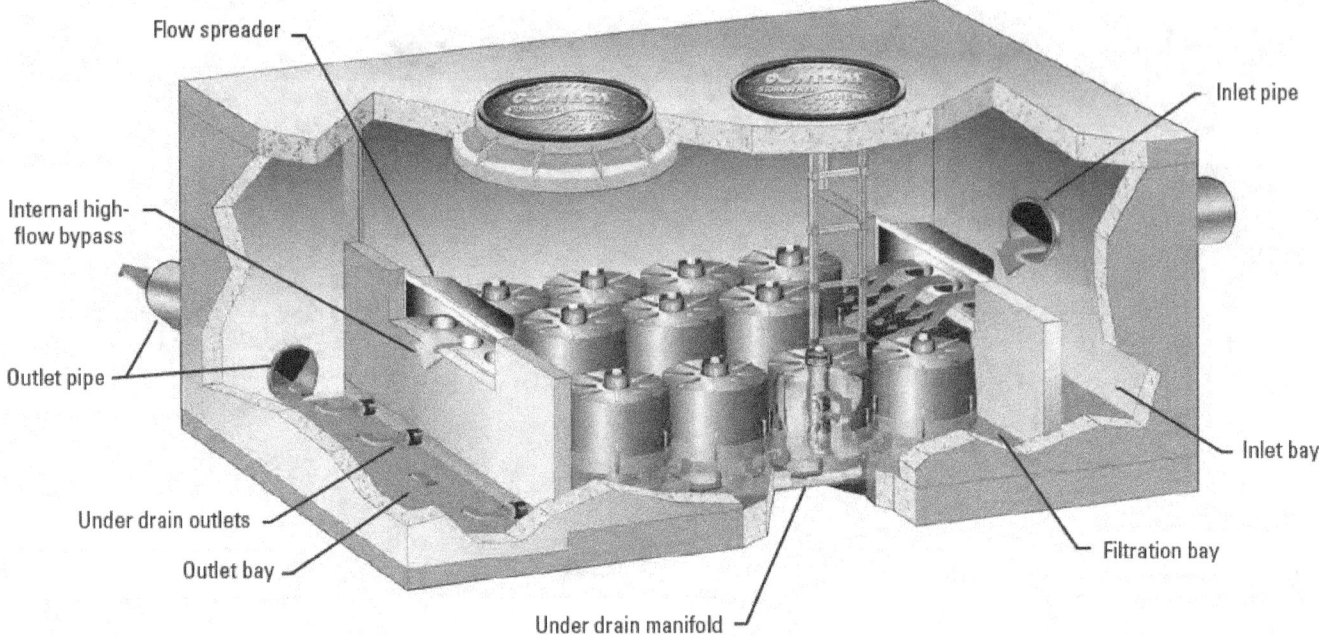

Figure 4.   Schematic diagram of the stormwater-filtration device (J.H. Lenhart, written comm., 2008). Image courtesy of Contech Construction Products Inc., used with permission.

is controlled through the cartridges by siphon action, and the water leaves the cartridge by an underdrain manifold. Each cartridge was designed to treat a peak flow of 0.033 ft³/s. The device was designed to treat flows with a peak flow rate of 0.297 ft³/s. When flows exceed 0.297 ft³/s, water bypasses the filter cartridges; at a height of 1.67 ft. stormwater goes over the high-flow bypass weir. Treated water from the underdrain manifold and untreated-bypass water enter into the outlet-bay area, which is 1.5-ft wide and 1.67-in. deep, then flows through an 8-in. pipe (U.S. Environmental Protection Agency, 2004). The SFD influent piping system was similar to that shown in figure 2.

Sampling Methods

Selection of sampling methods was based, as much as possible, on what has been learned from previous stormwater-monitoring projects in Wisconsin. Although methods for collecting precipitation, flow, and water-quality data have been used in previous Wisconsin projects, it still was important to perform quality-control tests and to make adjustments when problems were observed with the sampling methods. Extensive calibration efforts insured quality precipitation and flow data. Blanks and replicate samples were collected to evaluate the overall precision of the analyses. Some characteristics of the study sites, such as small-diameter pipes, high-flow velocities, and pipes with backwater flows, created complications with the sampling methods that had to be resolved during the project.

Measurement of Precipitation Depths

A tipping-bucket rain gage was used for continuous measurement of precipitation (fig. 5). The rain gage was located 25-ft northeast of the HSD monitoring station, attached to a barrier wall. It was mounted onto a 4-in. x 4-in. plank and raised 10 ft to avoid interference of nearby structures and prevent vandalism. A data logger recorded the number of bucket tips (0.01 in. per tip) every 60 seconds. This gage was not designed to record frozen precipitation, so values during periods of snowfall and freezing precipitation were not used. Calibration data indicated no need to adjust the original precipitation record, and the rain gage was cleaned during calibration. All precipitation data collected for each site are shown in appendix tables 2–1 and 3–1.

Calibration of Flow

Corrections were applied to stage measurements that reflect differences between water-surface elevations measured manually and those measured with the area-velocity flowmeters. To generate two sets of elevations for comparison, the pipe was first blocked by an inflated ball. Water levels were then increased in the pipe, and stage measurements were made at various levels representing the entire 8-in. depth of the pipe. Results from this procedure were used in making stage corrections throughout the entire period of record; accuracy of the record, on average, was estimated to be within ± 2 percent.

Flow calibrations were performed at the site on April 18 and Nov. 8, 2003. A 3-in. Parshall flume was mounted in a

Figure 5. Rain gage used for continuous measurement of precipitation.

level position in the back of a boom truck (fig. 6A). Water was pumped from the Milwaukee River into a 2.5-ft wide, 8-ft long, 2-ft deep chamber also mounted in the boom truck, just upstream from the flume, at four different pumping rates (fig. 6B). The pumping rates were approximately 0.1, 0.15, 0.4 and 0.55 ft³/s. Water levels in the flume (fig. 6C–D) were carefully monitored and recorded. The flow left the flume and passed over the flowmeters leading into the devices. Measurement of flow rates above 0.55 ft³/s was not possible owing to the chamber capacity and turbulence.

Several steps were taken to correct each area-velocity meter's record of flow. First, meters outputted point velocity that had to be converted to an average velocity by applying an equation supplied by the manufacturer. Overall, this conversion lowered flows by an average of 10 percent. Second, the cross-sectional area of the pipe was reduced to the area that effectively could carry flow by excluding the probe area and cord area in the flow calculation; this could be as much as one-half of the area at depths less than 0.1 ft. Flow was then calculated by multiplying the average velocity by the effective cross-sectional area.

Flow was estimated for each meter after stage corrections were applied. To determine a correction coefficient for the area-velocity meters, the corrected stage heights and flume flows were plotted. From this plot, a stage-flow rating was developed that passed through the flume-recorded points at stages ranging from 0.08 to 0.2 ft by using the USGS rating curve for stable channels. This equation is used to correct irregular channel flow at lower depths (equation 1). The meter acted as a control until a stage height of 0.08 ft was reached. Effective flow 'E' was set at 0.08 ft. The 'N' and 'C' coefficients from equation 1 were

adjusted to plot through the flume stage flow relation. The 'C' coefficient was based on the range of flows at a 0.8-ft depth, and 'N' fell into the ranges suggested by Rantz and others (1982). The flume rating was valid at depths ranging from 0.08 to 0.2 ft. A high-flow rating (depths greater than 0.2 ft) and a low-flow rating (depths less than 0.08 ft) were developed, by use of Manning's equation (equation 2), that matched the higher or lower end of the USGS rating curve.

The USGS rating curve for stable channels (equation 1) was used, which is a modified form of Manning's equation (Rantz and others, 1982)

$$Q = C \times (G - E)^N \qquad (1)$$

Where:
  Q is flow in cubic feet per second;
  C is flow coefficient;
  G is gage height of the water surface, in feet;
  E is effective zero control, in feet; and
  N is slope of the rating curve.

Flows were rated by use of Manning's equation (Rantz and others, 1982) when flume flow could not be used to rate the area-velocity meter. Also, calibration data were not available for low flows (less than 0.08 ft) because the area-velocity flowmeter does not register velocities at depths less than 0.01 ft. In addition, a second Manning's rating was used for flows greater than 0.20 ft because calibration data were not available beyond 0.2-ft stage, owing to the difficulty in maintaining laminar flow through the Parshall flume. Roughness coefficients were adjusted for high and low ratings that fit through the USGS rating curve. Roughness coefficients for the HSD were 0.0067 for the low rating and 0.0075 for the high rating, and roughness coefficients for the SFD were 0.0162 for the low rating and 0.0207 for the high rating.

The Manning's rating curve is provided in equation 2 for inch-pound units:

$$Q = (1.486)/n \times A \times R^{2/3} \times S^{1/2} \qquad (2)$$

Where
  Q is flow, in cubic feet per second;
  1.486 is a conversion factor to inch-pound;
  A is cross-sectional area in square feet, based on the water level;
  R is hydraulic radius, in feet, based on the water level;
  S is energy slope, in feet per foot; and
  n is Manning's roughness coefficient.

This method of estimating flows seems to be robust for volume relying only on the area-velocity flowmeter's stage. Results from this project show that calibration of the automatic flow-measurement equipment is critical for research projects of this type. Load estimate at the bottom of the device would be in error owing to the inaccuracies of flow measurement resulting in an over or underestimate of loads reduction for the device.

Figure 6.   Flow-calibration equipment: *A*, boom truck, *B*, top view of approach section of the flume, *C*, Parshall flume, and *D*, side view showing flow to the device.

Water-Quality Sampling

Automatic samplers (fig. 7) were used to collect flow-weighted samples at the inlet and outlet of each treatment device. The data loggers in the monitoring stations were programmed to initiate a subsample for a predefined volume of flow; consequently, more subsamples were collected for large-volume events than for small-volume events. In this respect, the sampling frequency increased or decreased to reflect the magnitude of flow. Flow-weighted sampling allowed for the collection of one composite runoff-event sample consisting of numerous subsamples throughout the course of the event. This approach resulted in a single average or "event-mean" concentration for each runoff event.

The intake of each inlet sampler was 3-ft upstream from each device, and the intake of the outlet sample line was 3-ft downstream from each device. The area-velocity flowmeters were 4-in. upstream from the sample intakes. All sample intakes were perpendicular to flow and approximately 1 in. off the bottom of the pipe. When a sample was initiated, the sampler went through a purge-and-rinse cycle before collecting the water-quality sample. This purge-and-rinse cycle was needed to eliminate residual water from the 3/8-in. Teflon-lined sample tubing.

The constituent list was based on the performance information from the manufacturers and the types of constituents WisDOT might want to regulate in the future (tables 1 and 2). Samples were analyzed at the Wisconsin State Laboratory

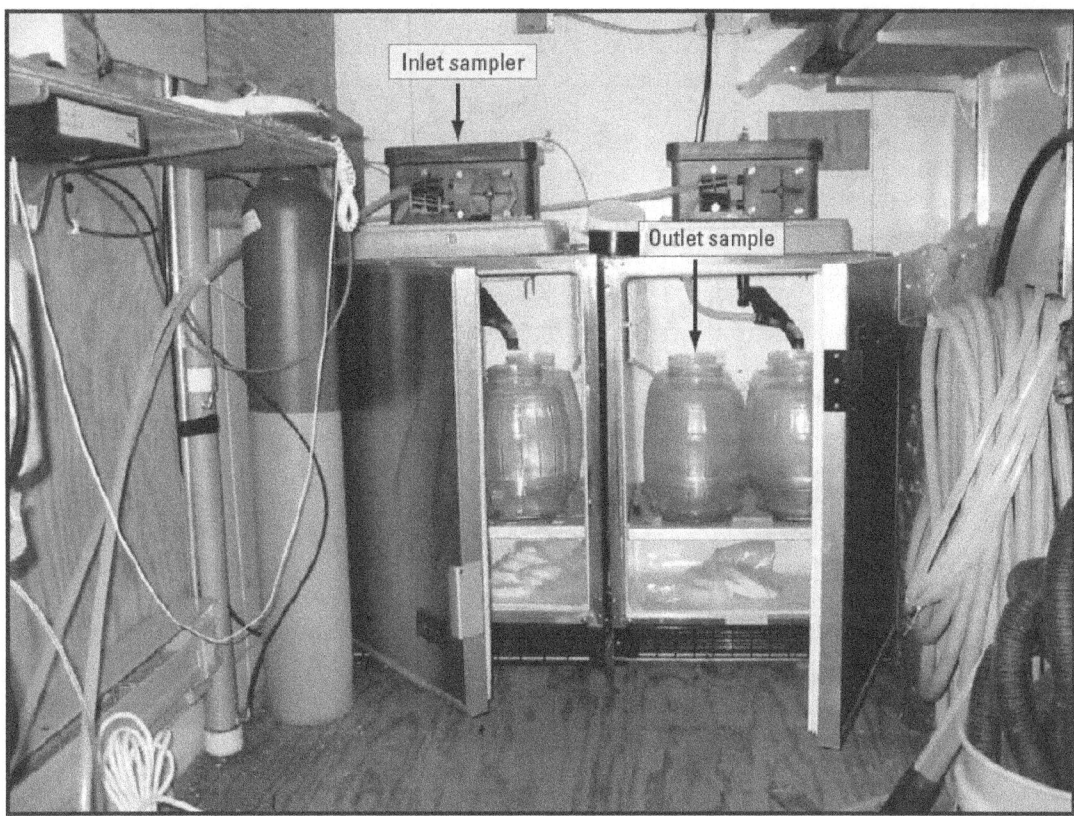

Figure 7.   Automatic water-quality sampling equipment.

of Hygiene (WSLOH). WSLOH is National Environmental Laboratory Accreditation Conference-certified and participates in the USGS Standard Reference Sample (SRS) program (Woodworth and Connor, 2003).

Analyses of suspended sediment (SS) and TSS are two different methods used for the determination of concentrations of solids. For the TSS method, an aliquot of a sample is filtered and weighed to determine the concentration of solids (Kopp and McKee, 1979). The method used to determine the concentration of SS requires filtering the entire sample (American Public Health Association and others, 1989). Concentrations of SS account for all of the solids within the sample and may yield higher concentrations of solids than those determined using the TSS method (Gray and others, 2000).

Particle-Size Analyses

Particle-size analyses of runoff-event samples were done in three different ways (table 1). The first level particle-size definition was the "sand/silt split," which was used

to determine the percentage of sediment, by mass, with a diameter greater than 62 µm (for simplicity, referred to hereafter as "sand") and less than 62 µm (referred to hereafter as "silt"). To define the sand fraction of the sample further, a visual-accumulation (VA) tube analysis was completed (Guy, 1977). This analysis determines the percentage of sediment, by mass, with diameters less than 1,000, 500, 250, 125, and 62 µm. To determine the silt fraction of the sample with more definition, a pipette analysis was done (Guy, 1977). This analysis determined the percentage of sediment, by mass, with diameters less than 31, 16, 8, 4, and 2 µm.

Monitoring Complications

For each device, the monitoring period was extended because of monitoring complications.

The HSD had four sets of problems:

Low-flow weir. In May 2002, it was noticed that the hydro-break or low-flow weir was not installed properly. The manufacturer replaced it with a 4-in. orifice plate.

Table 1. List of inorganic constituents analyzed, limit of detection, limit of quantification, and analytical method for samples collected at the hydrodynamic-settling and stormwater-filtration devices.

[mg/L, milligrams per liter; μg/L micrograms per liter; ASTM, American Society for Testing and Materials; EPA; U.S. Environmental Protection Agency; P, Phosphorus; SM, Standard Methods; VA, Visual accumulation tube; NA, not applicable]

Constituent or characteristic	Unit	Limit of detection	Limit of quantification	Method
Dissolved solids, total	mg/L	50	167	SM2540C[1]
Suspended solids, total	mg/L	2	7	EPA 160.2[2]
Suspended sediment, total	mg/L	.1	.05	ASTM D3977–97[1]
Chemical oxygen demand (COD)	mg/L	9	28	ASTM D1252–88(B)[1]
Dissolved phosphorus	mg/L as P	.005	.016	EPA 365.1[2]
Phosphorus, total recoverable	mg/L as P	.005	.016	EPA 365.1[2]
Calcium, total recoverable	mg/L	.02	.07	EPA 200.7[1]
Magnesium, total recoverable	mg/L	.03	.7	EPA 200.7[1]
Dissolved zinc	μg/L	16	50	EPA 200.9[1]
Zinc, total recoverable	μg/L	16	50	EPA 200.9[1]
Dissolved copper	μg/L	1	3	SM3113B[1]
Polycyclic aromatic hydrocarbon	μg/L	Varies	Varies	SM8310[1]
Copper, total recoverable	μg/L	1	3	SM3113B[1]
Sand/silt split and VA	NA	NA	NA	Guy (1977)
Five-point sedigraph (fall diameter)	NA	NA	NA	U.S. Geological Survey[3]
Sand fractionation	NA	NA	NA	Guy (1977)

[1] American Public Health Association and others (1989).

[2] Kopp and McKee (1979).

[3] Knott and others (1993).

Table 2. List of organic constituents analyzed, limit of detection, limit of quantification, and analytical methods for samples collected at the hydrodynamic-settling and stormwater-filtration devices.

[All data in micrograms per liter, determined by use of method SW8310 in American Public Health Association and others (1989)]

Constituent or characteristic	Limit of detection	Limit of quantification
1-Methylnaphthalene	0.046	0.14
2-Methylnaphthalene	.034	.11
Fluorene	.20	.65
Acenaphthene	.060	.19
Acenaphthylene	.072	.23
Anthracene	.021	.067
Benzo[a]anthracene	.062	.20
Benzo[a]pyrene	.070	.22
Benzo[b]fluoranthene	.11	.34
Benzo[g,h,i]perylene	.078	.25
Benzo[k]fluoranthene	.070	.22
Chrysene	.027	.087
Dibenzo[a,h]anthracene	.038	.12
Fluoranthene	.080	.25
Indeno[1,2,3-cd]pyrene	.12	.39
Phenanthrene	.040	.13
Pyrene	.070	.22
Naphthalene	.038	.12

Position of inlet pipe and flowmeter. When monitoring began in June 2002, the flowmeter in the inlet pipe was 3 ft from the device. Water elevations in the pipe were the same as in the swirl chamber, which created backwater conditions that allowed sediment to drop out into the pipe. Sediment covered the meter and produced errors in stage and flow. To alleviate this problem, a small check dam was placed upstream from the meter in an effort to cause the sediment deposition to occur ahead of the meter. However, velocities in the pipe sometimes were too great, and sediment moved past the dam, again covering the meter. It was decided to move the inlet meter farther upstream, out of backwater conditions. The most efficient alternative was to move the piping above ground. The new piping was designed to prevent turbulent flow and to match the existing pipe slope (appendix 2, fig. 2–1). This moved the meter about 12-ft upstream from the device (fig. 8). The new piping was installed in January 2003. For the 15 events sampled before this date, data are not reported herein because of their unreliability and the reduced sampling frequency.

Low flow at outlet. The outlet meter flow measurements were inaccurate as a result of low flow in the pipe. Composite sampling was based on inlet flow and outlet sample threshold, because of the difficulty in measuring flow at the outlet. This offset the outlet samples to about a minute after the inlet samples, so that approximately the same water was collected.

At the SFD site, the meters at the inlet and outlet were not changed; however, for five runoff events at the inlet and one runoff event at the outlet, there were velocity dropouts (the velocity dropped to zero) lasting 1 to 15 minutes during high flows. Flows during the dropouts were recorded as zero, and no samples were collected because the sampling routine was based on flow-proportional sampling. The dropouts may have resulted from larger, sand-size particles covering the meter; air entrainment disrupts the electrodes on the meter; or velocities exceeding the meter's measurement limits because of nearly pipefull conditions. These events were sampled, but analytical results are not reported herein because of inaccuracy of the flow data and samples were not obtained over the complete hydrograph. A hydrograph displaying velocity drops is footnoted in appendix 3, figure 3–1. In future projects of this type, use of an ultrasonic area-velocity flowmeter may eliminate velocity dropouts.

Quality Control

Equipment blank and replicate samples were collected at the inlet and outlet of both devices and analyzed for the same constituents as those from runoff-event samples. Blanks were collected at the beginning and midpoint of the project to validate clean-sampling procedures.

Replicate samples were done for several runoff events to quantify the variability and precision in sampling procedures. Analytical precision is a measurement of how much an individual measurement deviates from a mean of replicate measurements (Burton and Pitt, 2002). The relative percent difference (RPD) was calculated to evaluate precision in procedures after sample collection. The targets are set by the WSLOH.

The RPD equation is

$$\%RPD = [(X_1 - X_2) / \overline{X}] \times 100 \tag{3}$$

Where

X_1 is concentration of constituent in a sample,
X_2 is concentration of a constituent in replicate samples, and
$\overline{X}$ is mean value of X_1 and X_2.

Hydrodynamic-Settling Device

Two equipment-blank samples were collected between events 9 and 10 (blank 1) and events 30 and 31 (blank 2), respectively, to validate clean-sampling procedures. The blank 1 sample had detectable concentrations of dissolved copper (DCu) and chloride (Cl), but both concentrations were below the limit of quantification (LOQ) at the inlet and the outlet. The blank 2 inlet sample had detectable total copper (TCu) and DCu, but concentrations were below the LOQ. In the blank 2 outlet sample, chemical oxygen demand (COD) exceeded the LOQ, but additional quality-assurance/quality-control (QA/QC) samples collected directly from the sampler and from the jar of blank water were accidentally discarded; therefore, the particular piece of equipment that may have contributed to the detection could not be determined (table 2–2). A possible source of the COD in the second blank is the methanol used to rinse the 2.5-gal glass sample containers. Some of the methanol might have remained in the container after it was rinsed with distilled water. This problem requires further testing, but it is premature to discount all the COD values in this study until further testing is completed.

Replicate samples were collected during events 9, 18, and 42 to quantify variability in the sampling process. The RPD target for TSS was 30 percent or less; for metals, the RPD target was 25 percent or less (table 2–3). In replicates for events 18 and 42, the target of 25 percent was exceeded for TCu; in replicates for event 42, the RPD target for total zinc (TZn) was exceeded. Additionally, calcium (Ca) and magnesium (Mg) exceeded targets in events 9 and 42. For all of the dissolved constituents, a low RPD was reported, but high RPDs were reported for some of the particulate constituents. The high RPD for particulate-associated constituents might be explained by churn-splitting procedures, where precision is known to decline with increasing sediment concentration and particle sizes (Rickert, 1997). Since the end of the Riverwalk data collection, a new procedure of sieving samples before churning has been incorporated at the USGS Wisconsin Water Science Center (Selbig and Bannerman, 2007).

Figure 8.   Piping modification for the hydrodynamic-settling device.

Stormwater-Filtration Device

Three equipment-blank samples were collected: before event 1 (blank 1), before event 9 (blank 2), and before event 19 (blank 3). Blank 1 had detectable concentrations of Cl and Ca, but they were below the LOQ. Blank 2 had a detectable concentration of total phosphorus (TP) above the LOQ that was not in blank 3. Blank 3 had detectable concentrations of COD, DCu, TCu, and Cl but these were below the LOQ for the inlet and the outlet (table 3–2).

Replicate samples were collected during events 9, 14, 19, 26, and 28 to quantify variability in the sampling process. The RPD target for TSS was 30 percent; the target for metals—Cl, Ca, and Mg—was 25 percent (table 3–3). Replicate results for events 9, 14, and 28 exceeded the RPD target for TCu of 25 percent. Replicate results for events 9, 14, and 19 exceeded the RDP target for TZn of 25 percent; the RPD target for DCu was exceeded in event 28. The poor precision might have resulted from using the churn while splitting the sample (Selbig and Bannerman, 2007). As stated previously, procedures that involve sieving samples before churning have indicated an increase in precision.

Evaluation of the Hydrodynamic-Settling and Stormwater-Filtration Devices

Precipitation depths, flow, particle size, and water-quality data all were important in evaluation of the effectiveness of the two treatment devices. A comparison of measured-precipitation depths and long-term trends in precipitation depths assisted in evaluating whether the measured data were representative of precipitation patterns in Milwaukee. Precipitation data also were useful in checking the accuracy of the flow data. The flow data were needed to determine the volumes of runoff entering and leaving the treatment devices. Inlet and outlet pollutant loads calculated from volumes and water-quality concentrations were the basis for one of the methods used to determine the effectiveness of the two devices. A second method for evaluating the effectiveness of the devices was based on water-quality concentrations. The particle-size distributions (PSD) were used to analyze trends in the concentration data and the device effectiveness.

Precipitation Data

One rain gage was operated for both devices from June 21, 2002, until October 8, 2004 (tables 2–1 and 3–1). Precipitation data collected from June 21, 2002, until December 28, 2003 (18 months), was used for the evaluation of the SFD. The precipitation data collected from April 30, 2003, until October 8, 2004 (17 months), were used for evaluation of the HSD. Seven months of the rain-gage data overlapped

for the two devices. The largest precipitation depth with water-quality samples was 1.67 in. for the SFD and 1.75 in. for the HSD, whereas the smallest precipitation depth sampled for both devices was 0.07 in.

Data from two National Oceanic and Atmospheric Administration (NOAA) weather sites in the Milwaukee area were used to check the monthly precipitation depths recorded for this study for reasonableness. One NOAA site is General Mitchell International Airport (GMIA), about 10 mi south of the study site, and the other is Milwaukee Mount Mary College, about 10 mi west of the study site (National Oceanic and Atmospheric Administration, 2007a,b). The record at the GMIA sites was used to determine whether the sampled events would reasonably represent the long-term mix of precipitation depths observed in the Milwaukee area.

Monthly precipitation totals measured for the study sites compared well with the totals reported for the two NOAA sites (table 3). There was less than a 25-percent difference among the totals for 83 percent of the months. Months with larger differences generally were summer months when precipitation amounts can vary substantially over a distance as small as 10 mi, owing to a predominance of localized convective storms in the summer. All of the annual totals compared well among the NOAA sites and this study. Three of the six annual totals at the NOAA sites were almost identical to the study totals. The total precipitation for 2004 was 5.4 in. more than the long-term average precipitation, but the total precipitation for 2003 was about 12.8 in. less than the long-term average (table 3).

Although the total precipitation depths for 2003 and 2004 were not the same as the long-term average precipitation, the long-term distribution of rainfall depths measured in the Milwaukee area was comparable. However, precipitation can affect the performance of stormwater-treatment devices, and projects determining the treatment efficiency of devices would benefit by sampling a mix of precipitation depths and intensities. To assess the mix of precipitation depths during the study period compared to long-term precipitation patterns, the distribution of monitored precipitation depths from this study was compared to the historical (1949–92) distribution of precipitation depths from the NOAA GMIA site. Probability distributions for both data sets were constructed by use of the Weibull plotting position (Helsel and Hirsch, 1992). Precipitation amounts for individual runoff events were computed for both data sets. Precipitation depths greater than or equal to 0.07 in. (the minimum depth sampled during this project) were ranked from lowest to highest depth. A cumulative probability distribution was then computed for both data sets by use of equation 4:

$$P_R = i_R / (n+1) \tag{4}$$

where: R is precipitation event, P_R is probability of an event having a precipitation less than that of event, i_R is ranking of event R, and n is total number of events in the dataset.

Except for a moderate deviation for precipitation depths ranging from 0.65 to 0.9 in., the distribution of the sampled events was similar to the long-term distribution (fig. 9), indicating the data collected for the two devices represent a mix of precipitation characteristics for the Milwaukee area.

Number of Precipitation Events with Water-Quality Data

Water-quality samples were collected for a number of events at both devices (tables 2–1 and 3–1). In all, 45 water-quality event samples (47 percent of 109 runoff events) were available for inlet-to-outlet comparison for the HSD and 33 for the SFD (42 percent of 106 runoff events). These numbers do not represent the actual number of water-quality samples because some event samples were combined with preceding- and (or) subsequent-event subsamples into one composite sample. Subsamples were combined when the time between the ending of one precipitation event and the beginning of the next event was brief. Fifteen events had concurrent water-quality data available for comparison at both devices.

Most of the unsampled events (60 to 70 percent) for both devices had precipitation of less than 0.2-in. depth. Not many of the small runoff events were sampled because it takes about 0.08 ft of water to activate the flowmeter. Of the 33 water-quality samples collected for the SFD, 10 were precipitation depths of less than 0.2 in., and only 1 sample for the HSD was

runoff from a single precipitation depth of less than 0.2 in. For precipitation depths of 0.2 in. or greater, the percentage of runoff-events sampled increased to about 70 and 60 percent for the SFD and HSD, respectively.

Flow Data

Neither device has an external bypass-flow structure, so the volumes measured at the inlets should be the same as the outlet volumes. Flows at the inlet were selected to calculate the volumes for the HSD. Measurements made with the HSD outlet area-velocity flowmeter were not reliable because the flows were frequently too low to properly submerse the probe. Volumes for the SFD were calculated at the outlet. Although most of the flows were similar at the outlet and the inlet, the flows at the inlet were less reliable. During several large runoff events, the velocities at the inlet dropped to zero as the flows started to peak. The outlet flows, however, were reliable during high-flow events that caused velocity dropouts at the inlet (appendix 3, fig. 3–1).

Peak flows, percent-runoff coefficients, and volumes at the HSD inlet and the SFD outlet for sampled events are presented in tables 2–4 and 3–4, respectively. Only 4 of the 45 sampled events with water-quality data at the HSD site exceeded the design peak-flow rate of 1.6 ft^3/s. Exceeding the design peak flow at the HSD site should not reduce the amount of water treated because all the water goes into the treatment

Table 3. Comparison of monthly precipitation between the U.S. Geological Survey rain gage at the Riverwalk site and the National Oceanic and Atmospheric Administration weather sites at General Mitchell International Airport and Mount Mary College, Milwaukee, Wis.

[Precipitation is in inches; USGS, U.S. Geological Survey; GMIA, General Mitchell International Airport; MMMC, Milwaukee Mount Mary College; NOAA, National Oceanic and Atmospheric Administration 1997a, 1997b; — no data]

Month	USGS rain gage, water year 2002	NOAA GMIA, 2002	NOAA MMMC, 2002	USGS rain gage, water year 2003	NOAA GMIA, 2003	NOAA MMMC, 2003	USGS rain gage, water year 2004	NOAA GMIA, 2004	NOAA MMMC, 2004	NOAA MMMC, long-term averages
October	—	—	—	2.7	1.7	2.3	1.6	1.5	1.6	2.5
November	—	—	—	1.1	—	—	2.1	3.9	3.0	2.7
December	—	—	—	.9	.4	1.1	2.3	2.0	1.4	2.2
March	—	—	—	1.7	1.6	1.0	4.4	4.0	3.3	2.6
April	—	—	—	2.5	2.6	1.5	2.2	1.9	2.4	3.8
May	—	—	—	4.0	3.6	4.7	11.4	8.2	9.8	3.1
June	—	—	—	1.3	1.5	2.0	4.6	4.1	3.5	3.6
July	3.0	2.3	2.9	1.7	2.4	1.8	3.8	3.2	3.6	3.6
August	6.2	4.7	6.6	1.2	.6	1.4	4.1	3.4	2.6	4.0
September	3.6	2.8	3.3	1.5	1.6	1.9	.3	.2	.2	3.3
Total	12.8	9.8	12.8	18.6	18.6	17.7	36.8	36.5	31.4	31.4

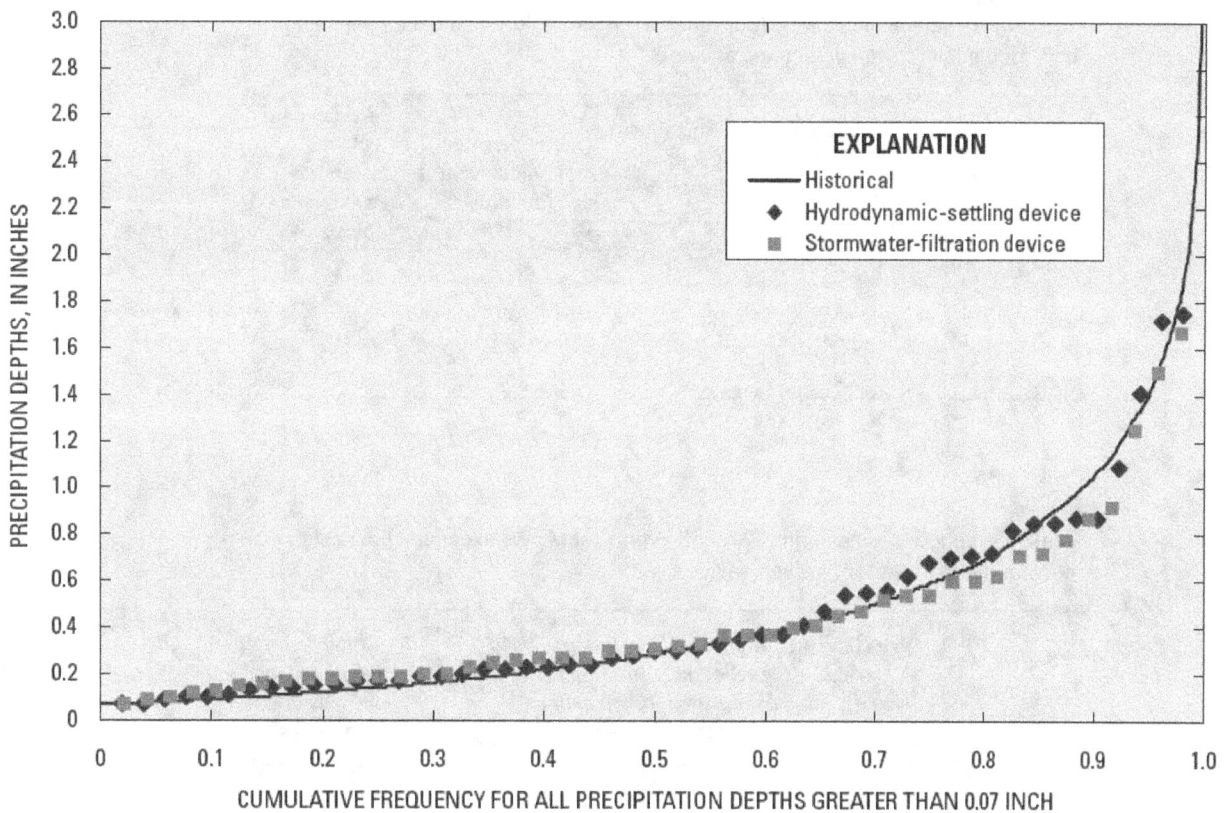

Figure 9. Cumulative precipitation distributions for the study period compared to historical precipitation records (1949–92) for Milwaukee, Wisconsin, based on the National Oceanic and Atmospheric Administration weather site at General Mitchell International Airport, Milwaukee, Wis.

chamber; however, flows greater than the maximum design peak flow exceed the optimal-treatment capacity for which the device has been sized. It was not possible to calculate the diminished treatment capacity for the few minutes the design flow was exceeded, because the sampling was done as flow composite (table 4). Nevertheless, because the design flows were exceeded only for a few minutes, the effect on the calculated loads should be minimal.

The design peak flow of 0.29 ft³/s was exceeded 12 times at the SFD site (table 5). Flow exceeded the design peak flow and the elevation of the bypass wall for sampled events 3 and 28; however, each time the design peak flow was exceeded or a bypass flow occurred, it only lasted for a few minutes. The treatment efficiency for each event does not appear to be affected, because most of the volume was treated below the design flow. Even if the bypass volumes had been sizeable, the efficiency calculations could have been done because the bypass water and treated water are mixed at the outlet.

Runoff coefficients can provide a check on the accuracy of the flow measurements (dividing the volume of precipitation into the runoff volume). A previous study on an elevated freeway indicated runoff coefficients near 85 percent (Pitt, 1987). Many of the runoff coefficients for the HSD and SFD sites were scattered near 85 percent (fig. 10). The numbers

of runoff events at the HSD and SFD sites with runoff coefficients of 100 percent or greater were 17 and 10, respectively. Only four runoff events with high runoff coefficients at the HSD site corresponded to flows exceeding the design peak-flow rate, whereas all the events with high runoff coefficients at the SFD site corresponded to flows exceeding the design peak-flow rate. Nine events at the HSD site had runoff coefficients of less than 50 percent, while five events at the SFD site had runoff coefficients of less than 50 percent.

Variability in the runoff coefficients could include the following:

Error in precipitation measurements—Error in precipitation measurements probably was a small role in the variability, since the comparisons with the local NOAA sites indicate that the precipitation data collected for the sites were reasonably accurate.

Uncertainties in the rating curves—A more appreciable source of the variability in the runoff coefficients could be the lack of high- and low-flow calibration data needed to extrapolate stage-flow rating curves. The uncertainty in the high-flow rating curves was more likely to be greater for events with higher peak-flow rates. High-flow rates were observed for all the runoff coefficients over 100 percent at the SFD site, and there was uncertainty in the low-flow rating curves for

Table 4. Length of time during four runoff events that flows exceeded the design flow for the hydrodynamic-settling device.

Date of event (event number)	Peak flow for event (cubic feet per second)	Time flow exceeded design flow (minutes)	Total duration of runoff event (minutes)	Percentage of time design flow was exceeded (percent)
9/14/03 (19)	2.08	2	412	1
5/21/04 (33)	1.81	3	67	4
6/14/04 (36)	2.64	5	47	11
8/03/04 (41)	2.44	11	230	5

Table 5. Length of time during 12 runoff events that flows exceeded the design flow for the stormwater-filtration device.

Date of event (event number)	Peak flow for event (cubic feet per second)	Time flow exceeded design flow (minutes)	Total duration of runoff event (minutes)	Percentage of time design flow was exceeded (percent)	Time flows exceeded the bypass wall (minutes)
06/21/02 (1)	1.11	6	46	13	0
07/08/02 (3)	1.06	22	145	15	9
08/21/02 (5)	1.12	11	985	1	0
09/02/02 (6)	.30	4	25	16	0
09/02/02 (7)	.38	5	264	2	0
06/08/03 (18)	.34	3	772	1	0
07/04/03 (20)	.36	3	2,302	1	0
07/21/03 (22)	.39	4	31	13	0
08/01/03 (24)	.33	1	3,552	1	0
08/25/03 (25)	.53	4	26	15	0
09/14/03 (28)	.52	6	412	1	4
11/04/03 (33)	1.12	5	196	3	0

events with small precipitation depths and long durations. The low-flow rating curve could result in an over or underestimate of the runoff volume for these small events, especially those event losses during small, long-duration events. An example of an overestimate is a runoff coefficient of more than 200 percent for a very small event at the HSD site (fig. 10).

Losses by traffic spray—Losses of water on the freeway surface also could contribute to the variability in the runoff coefficients. This is especially true for the lower runoff coefficients observed at both sites. One possible loss is the vehicles spraying the water over the sides of the elevated freeway. A second loss is evaporation and depression storage, especially for small, long-duration runoff events. Although these types of losses were not quantified, the potential for these

types of losses might mean the measured-runoff volumes are reasonably accurate for some of the events with low runoff coefficients.

Changes in the drainage areas—Changes in the size of the drainage areas might have contributed to the variability in the runoff coefficients. Each drainage area had a 0.5-percent slope, and the area was defined by the elevations above the inlet drains. The capacity of freeway drains may be exceeded during high-intensity precipitation events that produced an appreciable amount of runoff instantly; this could compromise the inlet drain diverting runoff from one inlet into another. The size of the drainage area could be decreased or increased by runoff bypassing the monitored inlets or runoff that bypassed upstream inlet drains into the monitored inlet.

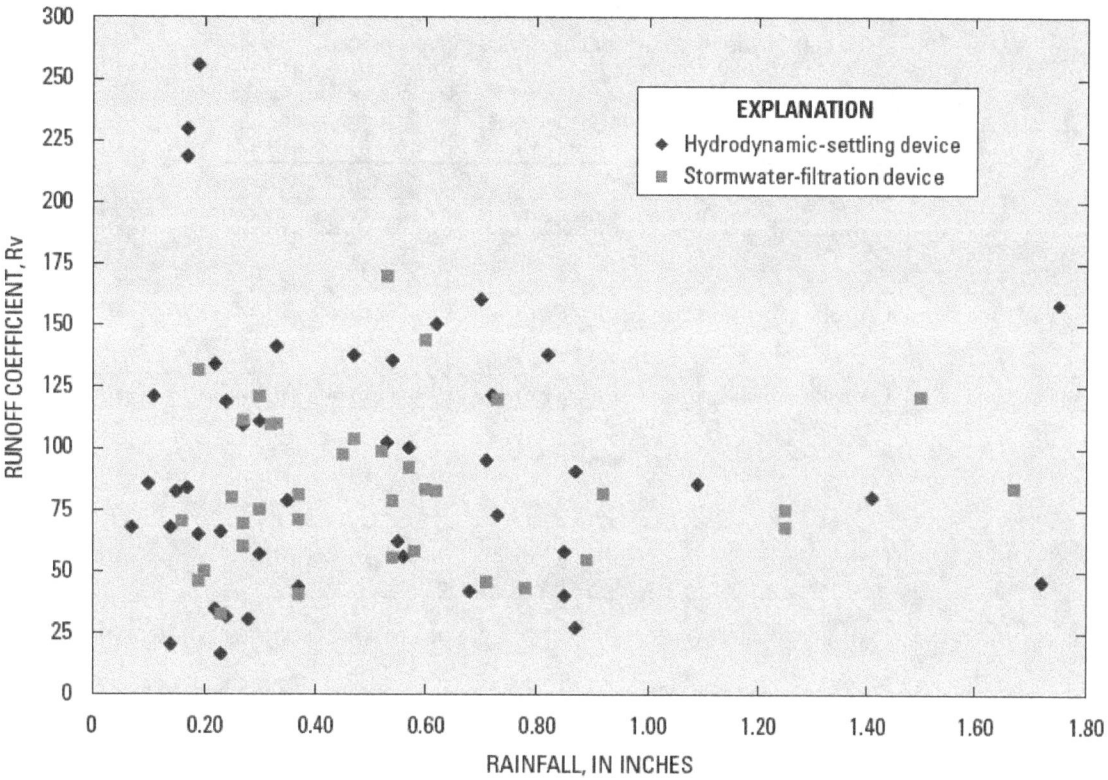

Figure 10. Freeway-runoff coefficients for the hydrodynamic-settling and stormwater-filtration devices.

Particle-Size Distributions

Particle-size distributions measured for this project could be helpful in designing devices to meet SS and TSS reduction goals for other elevated freeways. Proper selection of a PSD could be important for sizing a stormwater-control practice. Particle-size distribution with a large percentage of sand-size particles could support the selection of a smaller stormwater-control practice to achieve a reduction goal for SS or TSS, than would a distribution dominated by silt-size particles.

The PSD represents the concentration of SS in stormwater runoff because the particle-size analysis captures all the particles in a water sample. If the concentrations of SS and TSS are similar, the PSD also will represent the TSS. However, the PSD might not represent the distribution of other constituents, such as TP, because the concentrations tend to be higher on smaller particles (Dong and others, 1979).

Hydrodynamic-Settling Device

Sand/silt splits were collected for nine runoff events at the HSD inlet and outlet (table 6). Of those nine events, seven samples at the inlet had sufficient sediment content and sample volume for the VA-tube analysis. Three of the inlet

samples also had sufficient sediment content to do a complete PSD (table 7). Outlet samples contained enough sediment and sample volume for the VA-tube analysis in two samples and pipette analysis for one sample.

The sand/silt split at 62 μm (Guy, 1977) analysis shows that 8 of 11 inlet samples were composed mostly of silt or smaller particles (table 6). The average percentage of silt or smaller particles in the inlet samples was 67 percent, whereas the average for sand was only 33 percent. The PSD changed substantially from the inlet to the outlet of the HSD. A large proportion of the sand-size particles appear to be retained by the HSD.

The detailed particle-size results describe the relations between particle size and percent removal at the HSD site (fig. 11 and table 7). Based on the average PSD in these samples, if removal of all particles greater than 250 μm from the stormwater runoff averaged a 16-percent reduction in SS, then removal of all particles greater than 63 μm would result in an average 32 percent reduction in SS. To exceed an average removal of 32 percent at the HSD site, the device would need to retain some of the silt-size particles. For example, on average, the HSD would need to retain particles above 31 μm to achieve a 46-percent removal of SS. The levels of control for each particle size will vary somewhat with each event owing to the variability in the PSD among events.

Table 6.   Results of sand/silt split sediment analysis at the inlet and outlet of the hydrodynamic-settling device for nine events.

[μm, micrometer: %, percent by mass; ≥, greater than or equal to; <, less than; —, insufficient sample amount for determination of smaller particle size]

Event number	Inlet %		Outlet %	
	> 62 (μm)	< 62 (μm)	> 62 (μm)	< 62 (μm)
3	24	76	5	95
8	58	42	2	98
16	17	83	3	97
18	19	81	2	98
19	26	74	2	98
23	36	64	6	94
24	17	83	2	98
25	2	98	0	100
32	17	83	3	97
43	67	33	—	—
44	76	24	—	—
Median	24	76	2	98
Average	33	67	3	97
Maximum	76	98	6	100
Minimum	2	24	0	94

Table 7.   Particle-size distribution determined by visual accumulation (VA) and pipette-withdrawal analysis for seven events at the hydrodynamic-settling device inlet and outlet sampling sites.

[μm, micrometer; —, insufficient sample amount for determination of smaller particle size; all data are percent by mass; –, not analyzed; <, less than]

Particle size (μm)	Event number							Average
	3	16	18	25	32	43	44	
	Inlet							
<1,000	100	98	100	100	100	91	100	98
<500	96	96	100	99	98	85	84	94
<250	83	90	93	98	92	78	56	84
<125	79	86	85	98	86	48	35	74
<63	76	83	81	97	83	33	24	68
<31	—	74	74	—	—	—	15	54
<16	—	60	67	—	—	—	8	45
<8	—	45	55	—	—	—	5	35
<4	—	37	43	—	—	—	1	27
<2	—	29	28	—	—	—	1	19
	Outlet							
<1,000	—	100	—	—	—	—	—	—
<500	—	100	—	—	—	—	—	—
<250	—	100	—	—	—	—	—	—
<125	—	100	—	—	—	—	—	—
<63	—	100	—	—	—	—	—	—
<31	—	90	—	—	—	—	—	—
<16	—	80	—	—	—	—	—	—
<8	—	61	—	—	—	—	—	—
<4	—	51	—	—	—	—	—	—
<2	—	43	—	—	—	—	—	—

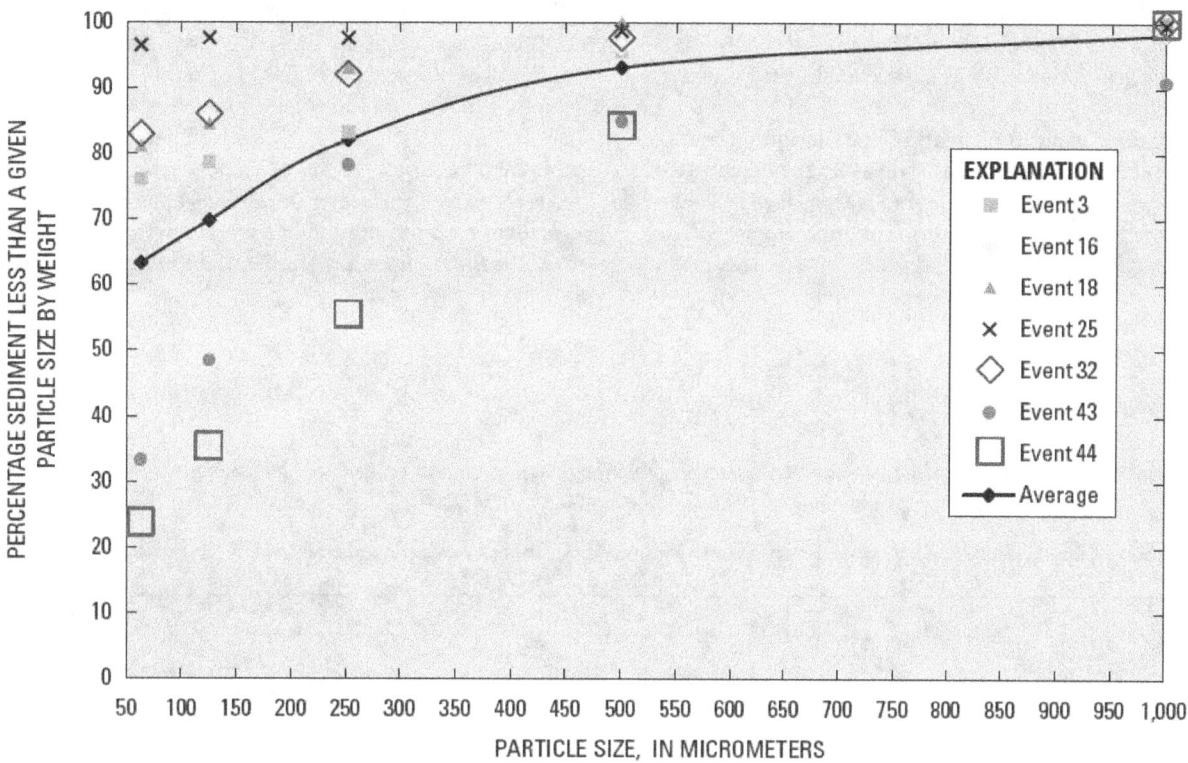

Figure 11. Particle-size distributions from the hydrodynamic-settling device inlet samples from seven events.

Stormwater-Filtration Device

Sand/silt splits were collected for 16 runoff events at the SFD inlet and outlet (table 8). Of those 16 events, 14 at the inlet had sufficient sediment content and sample volume for the VA-tube analysis (table 9). The VA-tube analysis could be done for only six samples at the outlet. Only 3 of the 16 events contained enough of the smaller-size particles for a pipette analysis at the inlet and outlet (table 9).

The sand/silt split analysis shows that 14 of the 16 inlet samples were composed mostly of sand particles (table 8). The average percentage of sand particles in the inlet samples was 71 percent, whereas the average for silt was only 29 percent. The PSDs changed substantially from the inlet to the outlet of the SFD. A large portion of the sand-size particles appeared to be retained by the SFD (table 8). The average percent sand at the inlet is 71 percent, and the average is reduced to 15 percent at the outlet.

There is a relation between detailed PSD and percent reduction in SS at the SFD site (table 9). Based on average PSD, removal of all particles greater than 250 μm would result in a 60-percent reduction in SS, and removal of all particles greater than 63 μm would result in nearly an 80-percent reduction in SS (fig. 12). The average percent removal for the 63-μm particles is larger in figure 12 than in table 8, because the two additional events in the sand/split data have much lower percentages of sand in the samples. Just the removal of

Table 8. Results of sand/silt split sediment analysis at the inlet and outlet of the stormwater-filtration device for 16 events.

[μm, micrometer: %, percent by mass; ≥, greater than or equal to; <, less than; —, insufficient sample amount for determination of smaller particle sizes]

Event number	Inlet %		Outlet %	
	> 62 (μm)	< 62 (μm)	> 62 (μm)	< 62 (μm)
1	82	18	9	91
3	88	12	12	88
4	77	23	6	94
5	68	32	18	82
7	92	8	8	92
9	68	32	9	91
10	72	28	0	100
11	60	40	0	100
12	66	34	0	100
14	70	30	0	100
15	85	15	56	44
24	8	92	2	98
25	11	89	4	96
26	96	4	99	1
28	87	13	8	92
30	100	0	—	—
Median	74	26	8	92
Average	71	29	15	85
Maximum	100	92	99	100
Minimum	8	0	0	1

sand appears to achieve a high level of removal at the SFD site. Given the variability in the PSDs between events, the levels of removal for each particle size would vary somewhat with each event.

Two similar sections of freeway in Milwaukee produced very different PSDs in the runoff. The average percentage of sand-size particles in the runoff samples from the HSD site was 30 percent, whereas the runoff samples from the SFD site contained 71-percent sand. This difference leaves much uncertainty in the selection of a PSD for freeways. Sand/silt splits also were determined for runoff samples collected from control and test sections from a freeway in another part of Milwaukee. The average percentages of sand-size particles calculated for the test and control sections were 46 and 34 percent, respectively (Waschbusch, 2003). Compared to the other three sites, the percent sand at the SFD site seems unusually high. It is possible that the difference in PSDs among the sites is more an artifact of the sampling than a real difference among the sites, depending upon the stratification of heavier material at the sampling location.

Table 9.   Particle-size distribution determined by visual accumulation (VA) and pipette-withdrawal analysis for 14 events at the stormwater-filtration device inlet and outlet sampling sites.

[μm, micrometer; —, insufficient sample amount for determination of smaller particle sizes; all data are in percent by mass; <, less than]

Particle size (μm)	Event number														Average
	1	3	4	5	7	9	10	11	12	14	15	26	28	30	
Inlet															
<1,000	80	52	84	100	71	93	93	90	86	90	92	90	72	78	84
<500	64	45	74	73	52	93	78	61	76	77	81	75	44	78	69
<250	36	25	38	42	17	58	40	47	54	49	34	23	23	67	40
<125	22	12	26	32	9	39	29	42	37	34	19	4	15	22	24
<63	18	12	23	32	8	32	28	40	34	30	15	4	13	0	21
<31	—	—	—	—	—	—	27	38	—	26	—	—	—	—	30
<16	—	—	—	—	—	—	26	33	—	20	—	—	—	—	26
<8	—	—	—	—	—	—	25	25	—	14	—	—	—	—	21
<4	—	—	—	—	—	—	23	16	—	11	—	—	—	—	17
<2	—	—	—	—	—	—	21	10	—	8	—	—	—	—	13
Outlet															
<1,000	100	100	—	—	—	—	—	100	100	100	100	—	—	—	100
<500	100	100	—	—	—	—	—	100	100	100	81	—	—	—	97
<250	98	100	—	—	—	—	—	100	100	100	57	—	—	—	93
<125	93	96	—	—	—	—	—	100	100	100	50	—	—	—	90
<63	91	88	—	—	—	—	—	100	100	100	44	—	—	—	87
<31	—	—	—	—	—	—	—	97	99	96	—	—	—	—	97
<16	—	—	—	—	—	—	—	96	93	86	—	—	—	—	92
<8	—	—	—	—	—	—	—	86	80	66	—	—	—	—	77
<4	—	—	—	—	—	—	—	78	61	55	—	—	—	—	65
<2	—	—	—	—	—	—	—	65	38	48	—	—	—	—	50

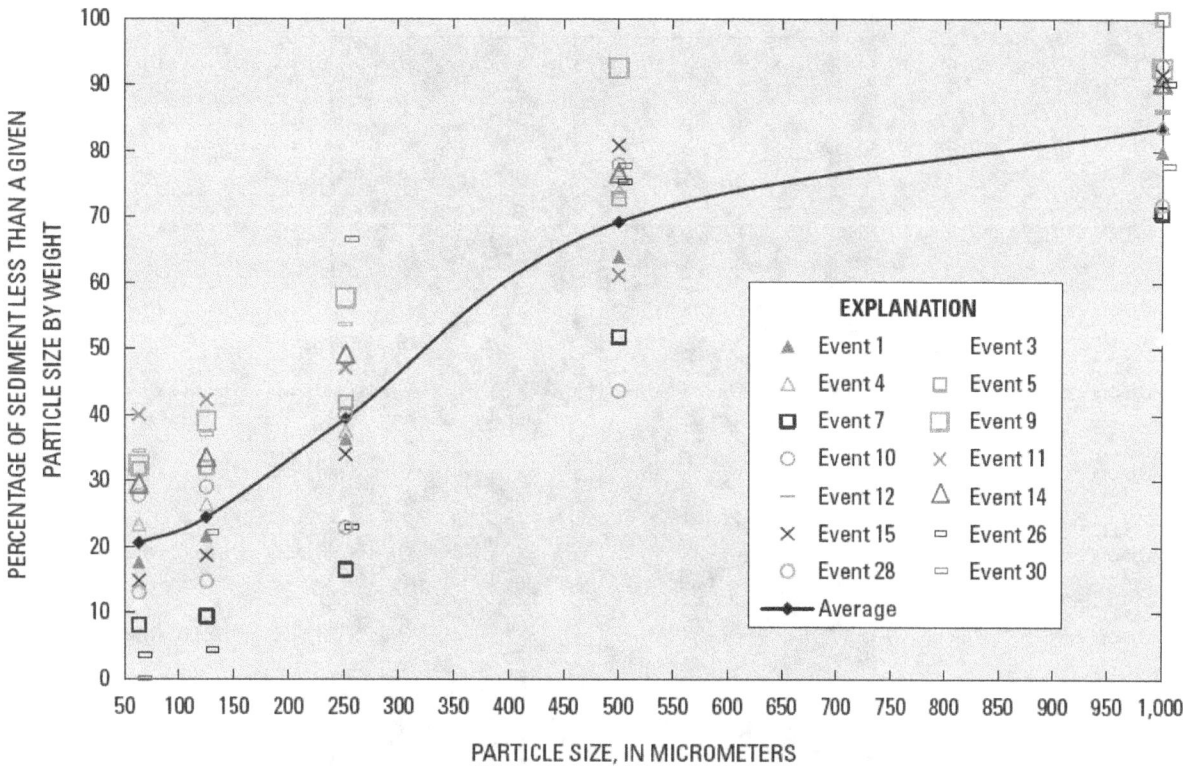

Figure 12.   Particle-size distributions from the stormwater-filtration device inlet samples from 14 events.

Summary of Inlet and Outlet of Chemical Concentrations

Chemical concentrations for each runoff event and the summary statistics for all the events, such as averages, medians, and coefficients of variations, are presented in appendix 2 for the HSD (tables 2–5 through 2–7) and appendix 3 for the SFD (tables 3–5 through 3–7). Thirty-two constituents were analyzed for the inlet and the outlet samples from both the HSD and the SFD (figs. 13 and 14). Eighteen of the constituents are different compounds of polycyclic aromatic hydrocarbons (PAHs). Samples from at least 15 runoff events were analyzed for all the constituents except PAHs. Samples from at least seven runoff events were analyzed for PAHs at both sites.

Runoff events with precipitation depths of less than 0.2 in. were analyzed for total dissolved solids (TDS), TSS, and SS, if adequate sample volume was collected. The number of events sampled was 33 at the SFD and 45 at the HSD. Samples were collected from June 21, 2002, to November 04, 2003, at the SFD site, and samples were collected from April 30, 2003, to November 15, 2004, at the HSD site. One sample at each site was collected during the winter.

Non-detectable concentrations were a substantial portion of the total for the PAHs; more concentrations were below detection limits for the outlets than the inlets. Five of the 18 PAH compounds that had large numbers of non-detectable concentrations were 1-methylnaphthalene, 2-methylnaphthalene, fluorene, acenaphthene, and acenaphthylene. To calculate

the summary statistics for total PAHs, a method was needed to fill in the non-detected concentrations. Summing the total PAH for an event-mean concentration (EMC) was done in one of three ways: (1) did not include non-detects, (2) used the detection limit for less-than detections, and (3) used one-half the limit of the detection value. The three summing methods resulted in means that were within ±5 percent of one-half of the applicable detection limit. To be consistent with other USGS studies, the total concentrations of PAH that had non-detects were not included in the summation (Mahler and others, 2005).

Most of the concentrations for the inlet and outlet follow a log-normal distribution. The Shapiro-Wilk statistic was used to test for normality (Helsel and Hirsch, 1992). The outlet concentrations for COD did not follow a log-normal distribution at the HSD site, nor did the outlet concentrations for Cl, COD, TSS, and TDS at the SFD site. Runoff data from highway sites around the country exhibit similar distributions for average concentrations; that is, they were either log-normal or can be approximated as log-normally distributed (Driscoll and others, 1990). The USEPA NURP study (U.S. Environmental Protection Agency, 1983) reached a similar conclusion for runoff-concentration data collected from many urban sites around the country. Data sets that are log-normally distributed are better described by the median or geometric mean than the arithmetic mean to reduce the affect of a few extreme observations; therefore, the medians and geometric means are listed in tables 2–5, 2–6, 3–5, and 3–6.

Figure 13.   Example of inlet and outlet hydrodynamic-settling device event samples.

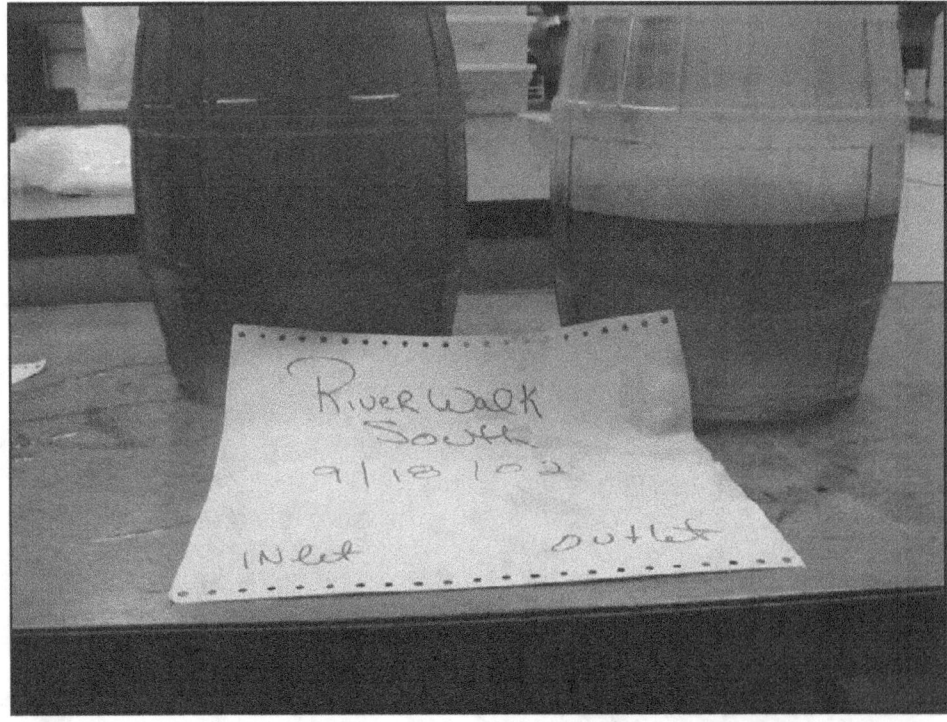

Figure 14.   Example of inlet and outlet stormwater-filtration device event samples.

Not all the same events were monitored at both sites, but the inlet medians for the HSD and SFD sites had similar constituents: dissolved phosphorus (DP), TP, DCu, and TZn (table 10). With differences in the range of 40 percent, the median inlet concentrations were relatively small among the PAHs, TSS, and TDS, but the SS median of 389 mg/L for the SFD (table 3–5) was about 3.4 times greater than the SS median of 114 mg/L for the HSD (table 2–5). Many individual event concentrations of SS observed at the SFD site were much higher than the observed concentrations of SS at the HSD site (fig. 15). It is expected that two similar sections of freeway might produce similar runoff concentrations for the constituents, but large differences in the concentrations of SS probably affected the differences in the PSDs between the two sites.

Although the inlet concentrations of SS at the SFD site were usually higher than the inlet concentrations of TSS, the inlet concentrations of SS and TSS were similar at the HSD site (fig. 16). The median inlet concentration of SS at the SFD site was 6 times the median concentration of TSS. The differences in the concentrations of SS and TSS might be explained by the possible exclusion of the larger sand particles from the TSS analysis at the SFD site. The dominance of smaller particles might explain the similarities in concentrations of SS and TSS at the HSD site. Gray and others (2000) observed that for several sites, the relations of concentrations of SS and TSS may be comparable when the percentages of sand-size material in the sample were less than 25 percent, but it cannot

be assumed that if there is no or very little sand the relation will exist.

The median outlet concentration of SS for the HSD was 67 mg/L, and the median outlet concentration of TSS was 47 mg/L (table 10). The SFD outlet medians for concentrations of SS and TSS were 34 and 36 mg/L, respectively. Outlet median concentrations of TP, DCu, dissolved zinc (DZn), and TZn were lower for the SFD than the HSD.

It is important to examine why two nearly identical source areas can have such large differences in concentration of SS. One possibility was that the SFD site had a source of larger particles, because more road-surface repairs were being done in the eastbound direction than in the westbound direction. Another explanation is that some of the sand-size particles were trapped somewhere in the pipe system before the runoff reached the HSD. Comparisons of events sampled on the same date from the SFD and the HSD substantiate that concentrations of SS were consistently higher at the SFD. There were two horizontal sections of pipe draining the westbound freeway into the HSD. These 8-in. pipes were beneath the freeway deck, about 15-ft above the ground. It is certainly possible that some of the larger material accumulated in this section of the pipe. Another possibility is that the difference in SS between the sites is more an artifact of the sampling than a real difference between the sites. The sample location of the HSD inlet was different from the SFD inlet (described in monitoring complications), which could affect sediment

Table 10.   Median concentrations at the inlets and outlets of the hydrodynamic-settling and stormwater-filtration devices.

[mg/L, milligrams per liter; µg/L, micrograms per liter; PAH, polycyclic aromatic hydrocarbons; <, less than]

Constituent	Hydrodynamic-settling device			Stormwater-filtration device		
	Number of samples	Inlet	Outlet	Number of samples	Inlet	Outlet
Dissolved solids, total (mg/L)	44	88	140	27	<50	78
Suspended solids, total (mg/L)	44	89	47	24	60	36
Suspended-sediment concentration (mg/L)	42	114	67	32	389	34
Chemical oxygen demand (mg/L)	18	63	75	17	51	50
Phosphorus, dissolved (mg/L)	18	.04	.028	17	.041	.037
Phosphorus, total recoverable (mg/L)	18	.146	.132	17	.152	.098
Copper, dissolved (µg/L)	18	14	15	16	14	12
Copper, total recoverable (µg/L)	18	53	41	17	44	23
Dissolved zinc (µg/L)	18	52	69	17	59	45
Zinc, total recoverable (µg/L)	18	231	172	17	226	91
Chloride, dissolved (mg/L)	18	20	38	15	9.2	17
Total PAHs (µg/L)	9	15	9	7	8	2

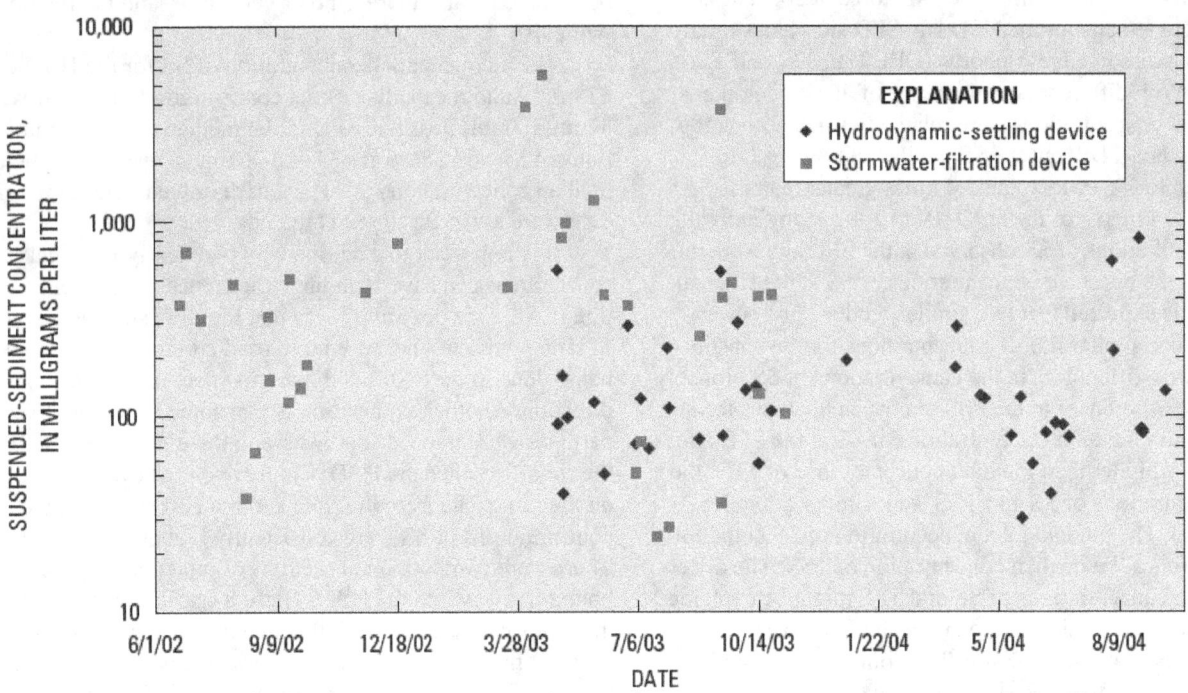

Figure 15.   Site comparison of suspended-sediment concentration from device inlets by date.

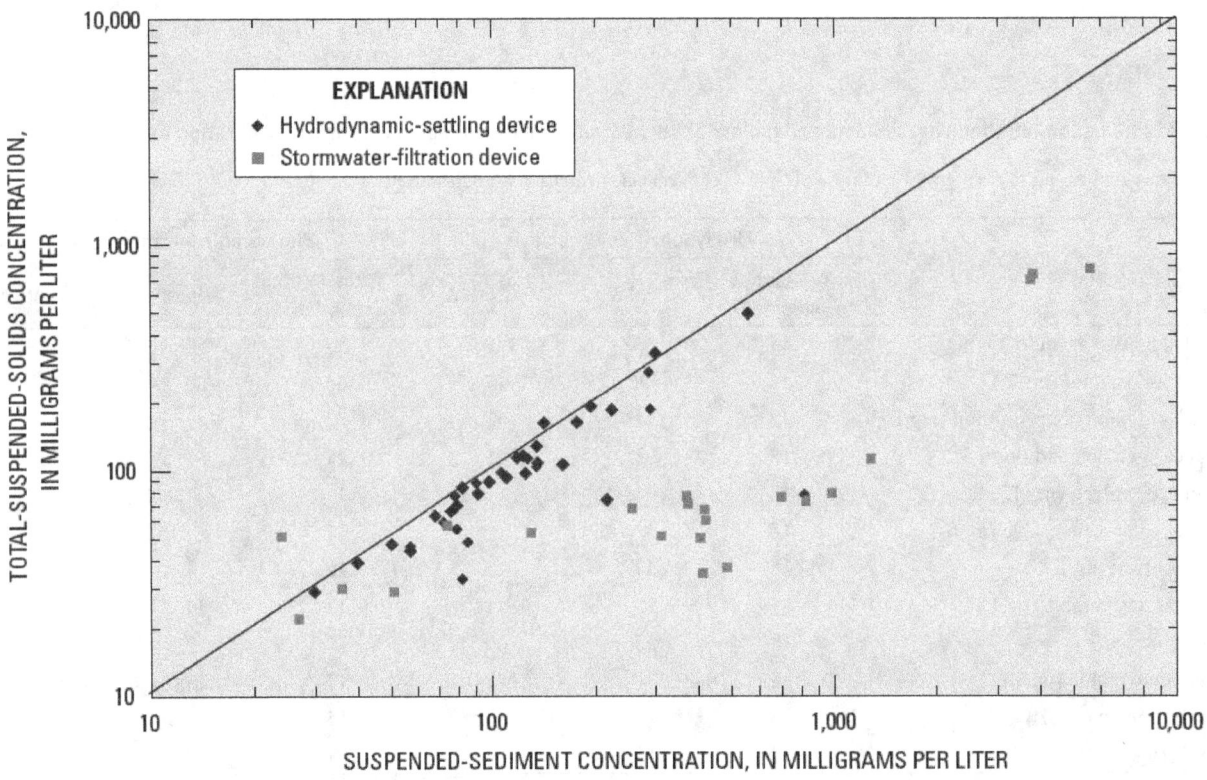

Figure 16.   Comparison of suspended-sediment concentration and total-suspended-solids concentration at inlet devices for the hydrodynamic-settling and stormwater-filtration devices.

Table 11. Comparison of mean influent concentrations at the hydrodynamic-settling and stormwater-filtration devices with average runoff concentrations from other highway sites.

[mg/L, milligrams per liter; HSD hydrodynamic-settling device; SFD, stormwater-filtration device; —, data not collected or not applicable; >, greater than; I, interstate]

Site	Percent impervious	Average daily traffic	Seasons sampled	Suspended solids, total (mg/L)	Chemical oxygen demand (mg/L)	Phosphorus, total (mg/L)	Zinc, total (mg/L)	Copper, total (mg/L)	Chloride (mg/L)
HSD	100	44,000	Nonwinter	117	78	0.18	0.25	0.07	27
SFD	100	44,000	Nonwinter	143	80	.20	.40	.10	59
I–794[1] Milwaukee	100	53,000	Nonwinter	138	105	.31	.35	.10	63
Multiple sites[2]	37–100	>30,000	Nonwinter	165	129	.52	.54	.06	31
I–894 National[3]	63	133,900	All	108	49	.10	.21	.06	511
I–894[3] Oklahoma (nonswept period)	94	133,900	All	197	49	.19	.32	.07	438
I–94[4] Minneapolis	55	114,000	All	118	207	.56	.17	.05	1,802
Arterial Street[5]	100	20,000	Nonwinter	241	—	.53	.55	.05	—
Highway[6] 12 &18 (beltline)	100	77,000	Nonwinter	106	—	.32	.125	.041	—

[1]Gupta and others (1981).

[2]Driscoll and others (1990); data from 12–16 sites.

[3]Waschbusch (2003).

[4]Thomson and others (1997).

[5]Bannerman and others (1992).

[6]Waschbusch (1996).

mixing through the pipes. Heavier material could have been stratified near the bottom of the pipe when it reached the SFD sampling tube, and because the tube is at the bottom of the pipe more of the heavier material was sampled.

Mean concentrations for the runoff samples from this project were comparable to the mean concentrations observed for runoff samples from other urban highways and arterial streets (table 11). Rural highways were not used for comparison, because the runoff quality was appreciably different among rural and urban highways (Driscoll and others, 1990). The concentrations of COD and TP were less than those from the other sites. Concentrations of TP should be lower because vegetated areas are an important source of phosphorus and the elevated freeway lacks vegetated areas. Chloride concentrations are highest for the sites with winter-runoff data. Data from the city arterial streets indicate that the results from this study also might apply to busy city streets. Comparable runoff quality among the urban highways increases the possibility that it is valid to extrapolate the concentration results from this study to other highways.

Efficiency Calculations

To determine the pollutant-removal efficiency of a stormwater-treatment device, various methods can be used (National Cooperative Highway Research Program, 2006). Two methods commonly used are the efficiency ratio and summation of loads (SOL) methods. The efficiency-ratio method

is defined in terms of the average of the EMC of pollutants over a period. The summation-of-loads method compares the efficiency of the summation of all inlet loads to the summation of all outlet loads.

Each method uses data from the inlet and outlet of the device to produce a single number that is designed to represent the pollutant-removal efficiency of the device; however, the methods do not evaluate whether there are statistical differences between the set of inlet and outlet concentrations. Therefore, it is important to supplement the efficiency calculations with a statistical test indicating whether the means of the concentrations are statistically different (Helsel and Hirsch, 1992).

A paired statistical test was used to determine whether the inlet concentrations were higher than the outlet concentrations. Most of the constituents were log-normally distributed; therefore, the nonparametric one-sided Wilcoxon signed-rank test was applied (Helsel and Hirsch, 1992). A test for significance was not done for Ca, Mg, and PAHs. Efficiency calculations were not done for Ca and Mg because they are used in the calculation of hardness. The small number of samples and the occurrence of censored data (values less than the detection limit) made it difficult to execute a significance test for the PAHs.

A paired-statistical test was considered valid for this data set because the inlet and outlet concentrations are paired for each event. It would be more difficult to defend the idea that the concentrations are paired if more of the outlet concentrations reflected the water stored in the devices between events.

Table 12. Summary of average of the event-mean concentrations and efficiency ratio as a percent for the hydrodynamic-settling device.

[PER, percent efficiency ratio; mg/L, milligrams per liter; µg/L, micrograms per liter]

Constituent	In	Out	PER
Dissolved solids, total (mg/L)[1]	213	627	−177
Suspended solids, total (mg/L)	117	67	42
Suspended sediment (mg/L)	170	73	57
Phosphorus, dissolved (mg/L)	.06	.04	28
Phosphorus, total (mg/L)	.18	.15	16
Copper, total recoverable (µg/L)	71	48	33
Zinc, dissolved (µg/L)	76	91	−19
Zinc, total recoverable (µg/L)	254	196	23
Chloride, dissolved (mg/L)	27	122	−347

[1]Kaplan-Meier analysis was used to compute summary statistics (Helsel, 2004).

For most events, the volume of inlet water was sufficient at the HSD site to replace the stored volume of about 30 ft³ at least 10 times. The same was true for the SFD site, where the stored water of about 20 ft³ was replaced at least 10 times during most events.

At the HSD, concentrations of SS, TSS, TP, DP, TCu, and TZn were significantly higher at the inlet than at the outlet at the 95-percent confidence level. Concentrations of three of the dissolved constituents—TDS, Cl, and DZn—were significantly lower at the inlet than at the outlet. There was no significant difference between the concentrations of COD and DCu.

Concentrations of 9 of the 11 constituents analyzed at the SFD site were significantly different at the 95-percent confidence level between inlet and outlet, and TP was significantly different at the 90-percent confidence level. Concentrations of DP were not significantly different at the inlet compared to the outlet. All the constituents that were significantly different were significantly higher at the inlet, except for Cl and TDS. They were significantly higher at the outlet.

Sufficient differences existed between the means of the inlet and outlet concentrations to have confidence in the efficiencies calculated for most constituents. Only the efficiencies for DP at the SFD site and COD and DCu at the HSD site were not considered significant.

Efficiency Ratio

The efficiency-ratio method of calculating efficiency of a treatment device weights all events equally. For example, a large-volume event with high concentrations will have the same weight as a small-volume event with low concentrations.

The calculation is represented by the following equation (U.S. Environmental Protection Agency, 1999)

$$\text{Efficiency ratio as a percent} = 100 \left[1 - \left(\text{Average of the outlet event-mean concentration} / \text{Average of the inlet event - Mean concentration} \right) \right] \quad (5)$$

Hydrodynamic-Settling Device

Of those constituents that were significantly different, six of the nine constituents had positive efficiency ratios at the HSD (table 12). Most of the efficiency ratios were about 30 percent or less. The TSS and SS efficiency ratios were higher, at 42 and 57, respectively. For Cl, TDS, and DZn, the negative efficiency ratios showed that the average event concentrations increased at the outlet of the HSD. Salt pellets from winter-road salting could have produced brine in the sedimentation chamber that increased the outlet concentrations for both Cl and TDS. For example, the inlet concentration of Cl on event 20 (4/17/04) was 40 mg/L, and the outlet concentration was 792 mg/L (table 2–6). It was not clear why the concentration of DZn increased at the outlet, but anoxic conditions from the stagnant water at the bottom of the device could have caused oxidation reduction of DZn-complexes.

The SS and TSS efficiency ratios for individual events (event efficiency ratio) tended to range from 50 to 90 percent when their inlet concentrations were greater than about 200 mg/L (fig. 17). Below this concentration, the event efficiency ratios were variable. When concentrations were around 150 mg/L, the SS and TSS event efficiency ratios ranged from 10 to 90 percent. An increase in efficiency ratios above 200 mg/L might be explained by a possible increase in the percentage of larger particles in samples with higher concentrations. This explanation for these tendencies cannot be tested when PSD data were available for only 8 of the 45 events monitored. Most the negative efficiencies were observed for inlet concentrations of SS and TSS of less than about 125 mg/L. The material retained at the bottom of the device was not at the manufacturer's prescribed level for a cleanout; therefore, negative efficiencies were more likely caused by scour and not because the device needed a maintenance cleaning.

At low peak flows, there was a wide scatter in the SS and TSS event efficiency ratios (fig. 18). For peak flows that were less than 0.15 ft³/s, the event efficiency ratios ranged from zero to nearly 100 percent. Most of the events were in this area of wide scatter. The peak flows were not a predictor of efficiency ratios except when peak flows were greater than the design flow. When peak flows were greater than the design flow, all of the event efficiency ratios were negative, except for SS in one event. Flows greater than the design flows were more likely to scour some of the sediment already retained in the device, and the amount removed could be more than that retained during the event. The HSD had no bypass, so all runoff entered the settling chamber of the HSD.

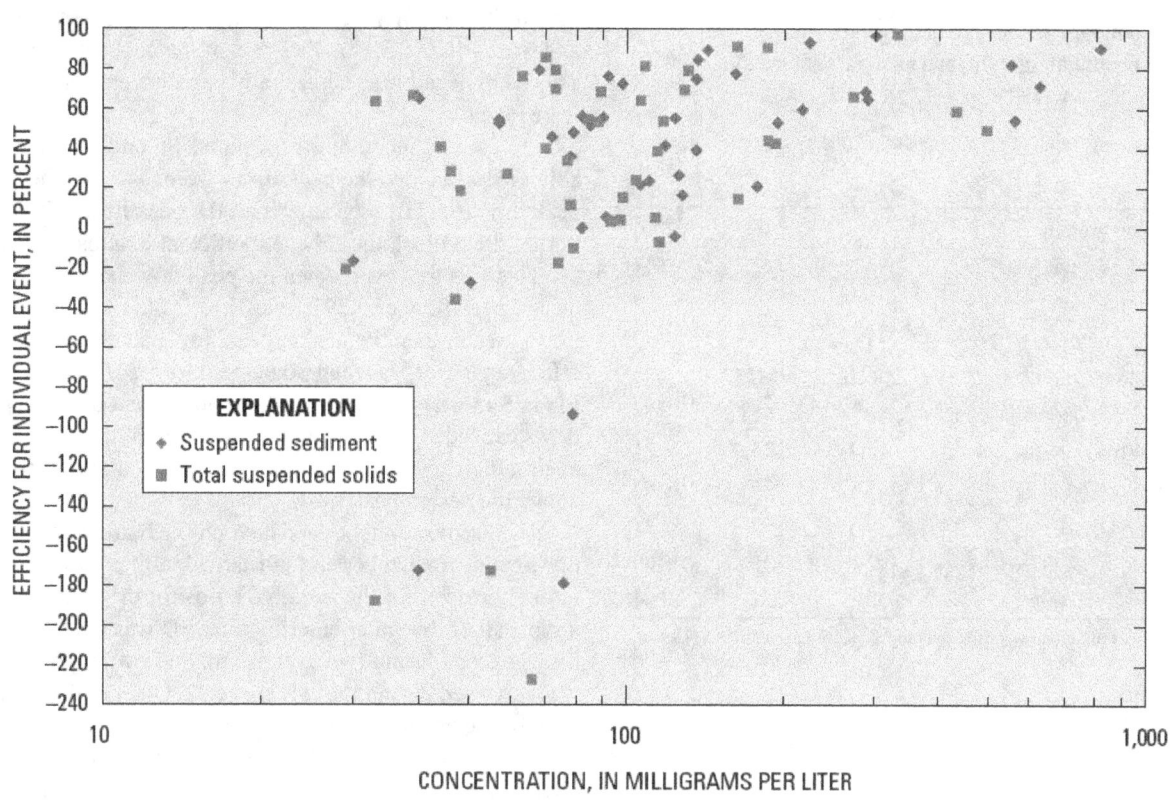

Figure 17. Efficiency ratios for total suspended solids and suspended sediment as a function of concentration for the hydrodynamic-settling device.

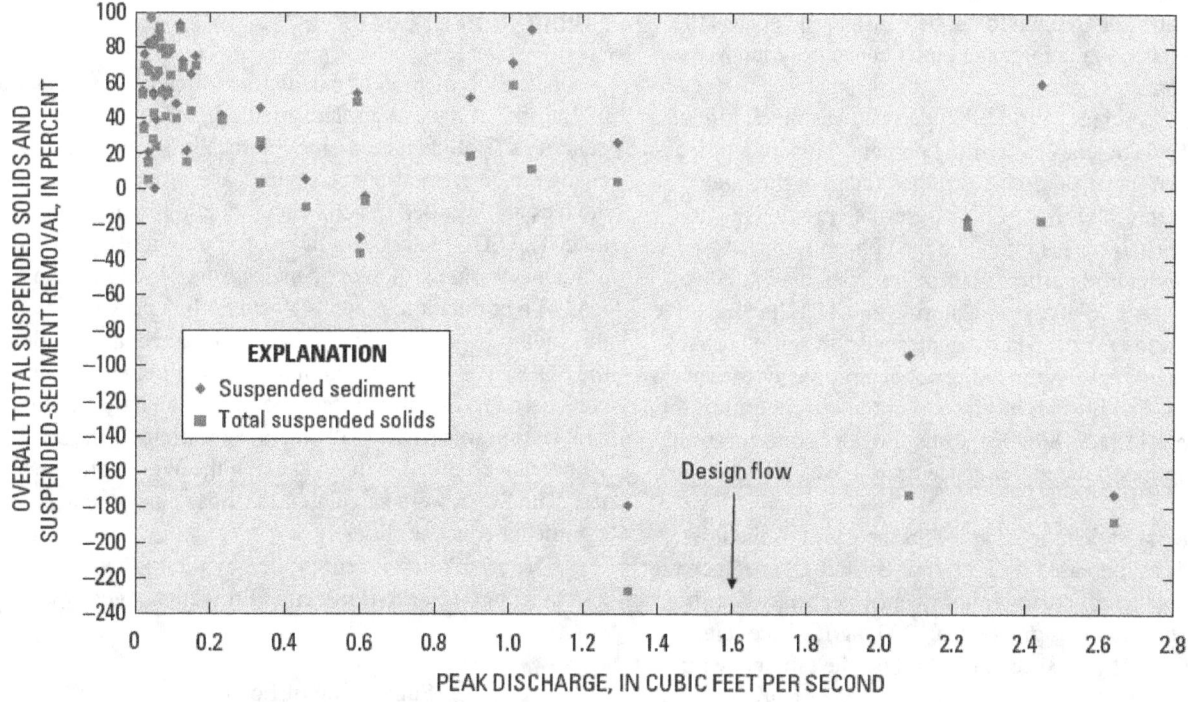

Figure 18. Removal efficiency of total suspended solids and suspended sediment as a function of peak flow for the hydrodynamic-settling device.

Table 13. Summary of average of the event mean concentrations and efficiency ratio as a percent for the stormwater-filtration device.

[%, percent efficiency ratio; mg/L, milligrams per liter; μg/L, micrograms per liter]

Constituents	In	Out	%
Dissolved solids, total (mg/L)[1]	98	262	–167
Suspended solids, total (mg/L)	143	58	59
Suspended sediment (mg/L)	743	73	90
Phosphorus, total (mg/L)	0.20	0.12	40
Chemical oxygen demand (mg/L)	80	65	18
Copper, dissolved (μg/L)	18.3	15.5	21
Copper, total (μg/L)	103	35	66
Zinc, dissolved (μg/L)	74	57	23
Zinc, total (μg/L)	402	135	66
Chloride, dissolved (mg/L)	59	231	–294

[1]Kaplan-Meier analysis was used to compute summary statistics (Helsel, 2004).

Stormwater-Filtration Device

Eight of the ten constituents had positive efficiency ratios for the SFD (table 13). All the constituents associated with TSS, such as TP and the total recoverable metals, had efficiency ratios ranging from 40 to 66 percent. These percentages are similar to the efficiency ratio of 59 percent for TSS. The efficiency ratio for DP should not be considered because the difference between the inlet and outlet concentrations was not significant.

The SS efficiency ratio of 90 percent was much higher than the TSS efficiency ratio of 59 percent. This may reflect the large amount of sand-size particles found in the inlet samples for the SFD (fig. 12). As described previously, the sand-size particles are included in the concentration of SS analysis, but a large portion of these particles may not be included in the TSS analysis. Chloride and TDS had negative efficiency ratios for the SFD. Again, road-salt brine probably affected the increase in the outlet concentrations of these two constituents. The biggest increase in outlet concentration for Cl was event 11 (12/18/2002), when the inlet concentration was 310 mg/L and the outlet concentration was 2,590 mg/L (table 3–6).

Efficiency ratios for SS and TSS increased as the inlet concentrations increased (fig. 19). For SS and TSS, 95 percent of the events had efficiency ratios over 70 percent when the inlet concentration was greater than 200 mg/L. Once inlet concentrations of SS exceeded 600 mg/L, the efficiency ratios were always about 90 percent. The presence of a greater number of large particles at the higher concentrations probably contributed to the consistently higher efficiency ratios at higher concentrations. Events with concentrations below about

120 mg/L had efficiency ratios ranging from 308 to 71 for SS and from –102 to 62 for TSS. Six of the concentrations of TSS were negative, whereas only two concentrations of SS were negative.

Event efficiency ratios for both TSS and SS were reasonably constant when the peak flows were less than the design peak flow (fig. 20). For almost all the peak flows observed during the project, the SS event efficiency ratios were above 60. This was true even when the peak flow exceeded the design peak flow of 0.297 ft³/s. Three of the events with peak flows greater than the design peak flow had either negative or efficiency ratios less than 30 percent for SS. Efficiency ratios for TSS mostly ranged from 40 to 60 percent until the peak flows exceeded the design peak flow. Six of the twelve TSS event efficiency ratios were negative for peak flows that were above the design peak flow.

Scour of sediment deposited on the bottom of the pre-treatment chamber and the cartridge-filter bay seems an unlikely reason for the negative ratios for the SFD, because almost all of the water entering the SFD was treated by the filters except during two events. Only a few minutes of bypass was observed during these two events. One possible explanation for the negative ratios might be the remobilization of the clay-size particles and very-fine or silt-size grain particles already trapped in the filters (J.T. Doerfer, oral commun., 2008). This is more likely to happen during the first flush of water into the filters after the filters have had a chance to dry between events.

Summation of Loads

The SOL method of calculating efficiencies is weighted by the size of the events. This method puts an emphasis on the quantity of pollutants entering the receiving water instead of a change in concentrations. In many cases, filtration or settling devices are installed to achieve a reduction in the pollutant load. The SOL method might be of more interest than the efficiency-ratio method to agencies trying to reduce the total load of a pollutant to receiving waters. It is possible, by use of this method, that a small number of large events can dominate the SOLs. The water stored in the devices between events was considered too small to affect the SOL calculations. Significant testing was performed on concentrations at the 95- or 90-percent confidence interval using the Wilcoxon signed-rank test. The SOLs were calculated for those constituents that had significant concentrations.

The calculation is represented by the following equation as a percent (U.S. Environmental Protection Agency, 1999):

$$\text{Summation of Loads} = 100 \times [1- (\text{Sum of Loads}_{Outlet}/\text{Sum of Loads}_{Inlet})] \qquad (6)$$

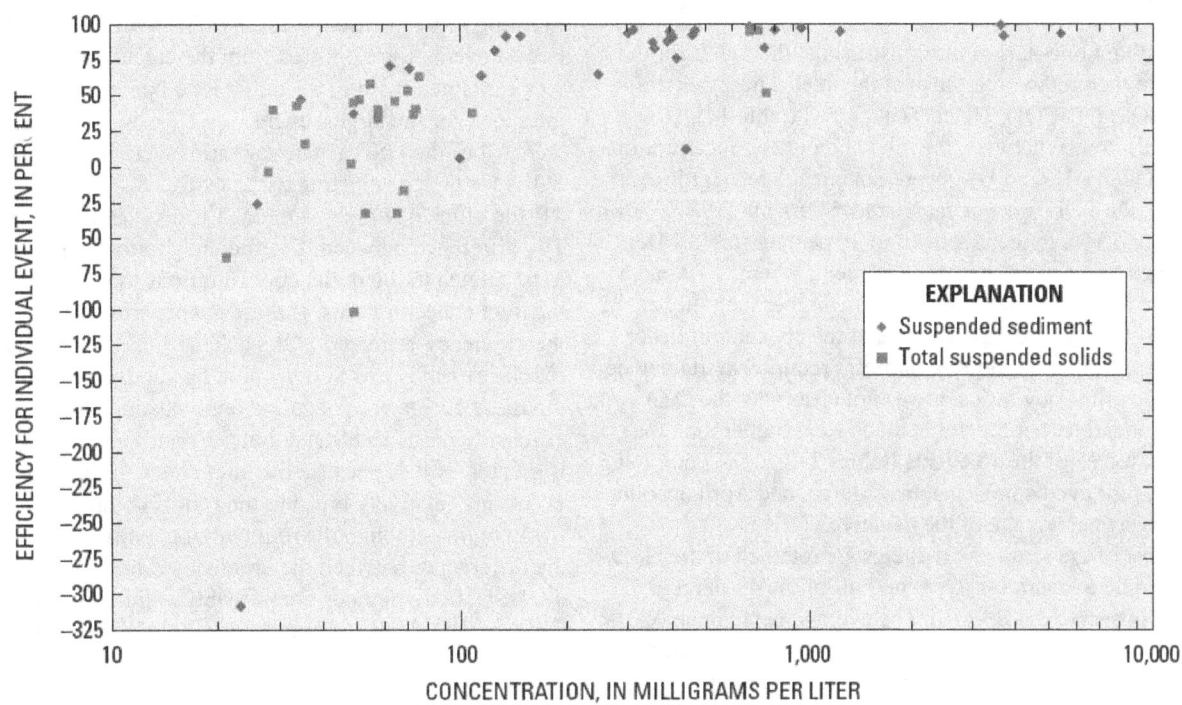

Figure 19.   Removal efficiency of total suspended solids and suspended sediment as a function of concentration for the stormwater-filtration device.

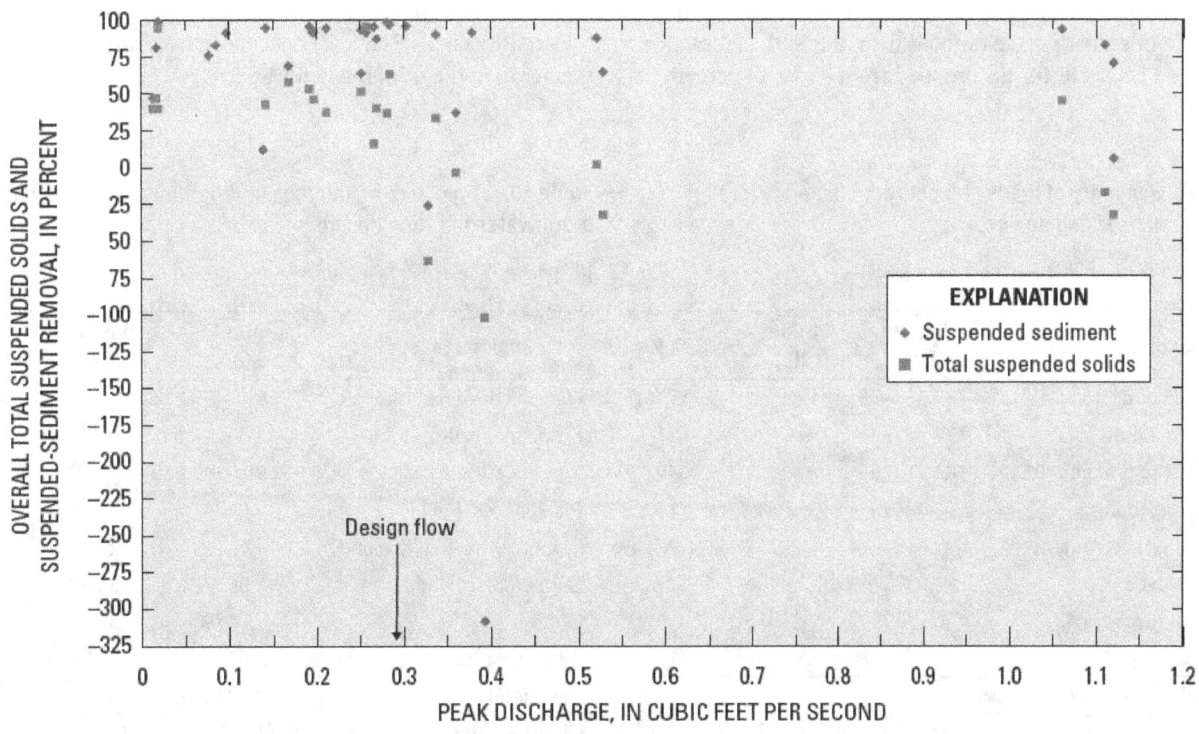

Figure 20.   Efficiency ratios of total suspended solids and suspended sediment as a function of peak flow for the stormwater-filtration device.

Hydrodynamic-Settling Device

For about one-half of the constituents, the SOLs for the HSD are higher at the outlet than at the inlet. These constituents are TDS, DP, COD, DCu, DZn, and Cl (table 14). However, the difference between the inlet and outlet concentrations for DP, COD, PAH, and DCu were not considered significant Therefore, the SOL was not calculated. SOLs for TP, TZn, TCu, SS, and TSS ranged from 10 to 49 percent (table 14). SS had the largest reduction, at 49 percent, whereas TP had a 10-percent reduction. SOLs for SS were higher than those for TSS because the laboratory analysis of the concentrations of SS may represent the sand-size particles retained in the device. As with the efficiency ratios, it was not clear why the DZn SOL was negative, but the outlet loads were higher than the inlet loads for 12 of the 18 events (tables 2–8, 2–9). For Cl at the outlet, four events in December, March, and April account for the largest percentage of the negative SOLs.

If most of the sand-size particles are retained in the HSD, the sand could account for over one-half of the 49-percent reduction observed for SS load. The average percentage of particles in the sand fraction entering the HSD was 33 percent (table 6); however, to bring the SS efficiency load up to 49 percent, some of the silt-size particles also must be trapped in the HSD. If sufficient water-quality particle-size data were available, correlation between the SOL and particle size may be possible.

SOLs are weighted by the size of the events, so a large event can have a disproportionate affect on the final efficiency loads. To test the effect of larger events, the events with the two largest inlet loads were omitted from the SOL calculation for TSS and SS. SOL for SS dropped from 49 to 40 percent,

about a 20-percent decrease without the two largest events. SOL for TSS increased from 25 to 30 percent when the two largest events were omitted from the calculation. This increase was explained by the TSS-outlet load being larger than the inlet load for the largest event.

All of the HSD efficiency ratios were higher than the SOLs for the constituents with positive SOLs. For TSS and TP, the efficiency ratios approach was twice that of the SOLs. The difference between the efficiency ratios and SOLs seems to be related to the outlet concentrations that were higher than the inlet concentrations. If these events are removed from the efficiency-ratio and SOLs calculations, the TSS and SS efficiency ratios and SOLs would be almost the same. Removing these events with negative removals increases both the efficiency ratios and SOLs, but the relative increase for SOLs is greater. This is because the inlet concentrations for these events are relatively low, but the runoff volumes are relatively high. Although the removal of negative events explains the differences between the efficiency ratios and SOLs for the HSD, both methods are probably valid depending upon whether the goal is to control concentrations or loads.

Stormwater-Filtration Device

All the SOLs, except for DP, were significant for the SFD. The highest SOL was SS, at 89 percent, whereas the rest of the positive SOLs ranged from 20 to 64 percent (table 15). SOLs for TSS, DZn, and DCu might have been somewhat higher if a few of the outlet loads were not greater than the inlet loads (tables 3–8 and 3–9). Only two outlet loads were greater than the inlet loads for SS.

Table 14. Summary of loads and percent efficiency for the hydrodynamic-settling device.

[lb, pound; %, percent; SOL, summation of loads]

Constituent load	Inlet (lb)	Outlet (lb)	SOL %
Dissolved solids, total[1]	143	417	−177
Suspended solids, total	127	94	25
Suspended sediment, total	182	92	49
Phosphorus, total	.0943	.0847	10
Copper, total recoverable	.0354	.0259	27
Zinc, dissolved	.0378	.0466	−23
Zinc, total recoverable	.1249	.1055	16
Chloride, dissolved	11	35	−218

[1]Load does not included non-detects.

Table 15. Summary of loadings and percent efficiency for the stormwater-filtration device.

[lb, pound; %, percent; SOL, summation of loads]

Constituent load	Inlet (lb)	Outlet (lb)	SOL (%)
Dissolved solids, total[1]	30	64	−112
Suspended solids, total	52	26	50
Suspended sediment	368	40	89
Phosphorus, total	.0624	.0391	37
Chemical oxygen demand	24.6	19.5	21
Copper, dissolved	.0052	.0042	19
Copper, total	.0279	.0111	60
Zinc, dissolved	.0242	.0194	20
Zinc, total	.1244	.0438	65
Chloride, dissolved	14	33	−136

[1]Load does not included non-detects.

The SOL for TSS was 50 percent, which was lower than that for SS. The reason for the difference was that inlet loads for SS were larger than those for TSS, but outlet loads were similar for most events (table 3–8). For example, the outlet loads for TSS and SS were the same on event 14, but the inlet load for SS was about 7 times the TSS load. The concentration of SS analysis includes all the larger particles that had major effects on the SOLs. This was especially true at the SFD site because such a large proportion of the inlet particles were in the sand fraction (table 8).

Based on the average percentage of the sand-size particles in the inlet runoff (table 8), total removal of the sand fraction would achieve a SOL for SS of about 70 percent. Some percentage of silt-size particles also must have been removed, because total removal of the sand-size particles would not cause the SOL for SS to be greater than 70 percent. The three events with pipette data for the outlet showed the SFD can remove some of the particles ranging from 8 to 16 µm (table 9); unfortunately, there was not enough particle-size data in this study to calculate a SOL by particle size. An ETV evaluation of an SFD in Georgia found a SOL by particle size was a 40-percent reduction in particles at the silt- and smaller-size fraction (U.S. Environmental Protection Agency, 2005a). If the SOL for the silt-size fraction was 40 percent at the Milwaukee site, the silt would account for about 12 percent of the SOL for TSS and SS. SOLs for TSS would be affected by the 12 percent for silt, the percentage of sand trapped by the HSD, and the ability of the TSS analysis to capture all the sand-size particles in the sample.

SOLs for TP and total metals ranged from 38 to 68 percent (table 15). Factors that might contribute to the high SOLs for these constituents are (1) none of the outlet loads were higher than the inlet loads (table 3–9), (2) some of the silt or smaller particles were captured, and (3) an appreciable portion of the dissolved metals were removed. The SOLs for DZn and DCu were about 20 percent. Removing the silt-size particles can increase the removal of TP and metals, because those concentrations tend to be highest on silt-size particles (Dong and others, 1979).

Removing events 13 and 14, which had the largest SOLs for TSS, decreased the SOLs for TSS from 52 to 37 percent. Removing these same two events from the SOL for SS only reduced the SOL for SS from 89 to 87 percent.

Efficiency ratios and SOLs were almost identical for the SFD. This is possible if the events with the higher concentrations also tended to have higher loads. These events would have a similar effect on the final SOLs and efficiency ratios. For example, 4 events—13, 14, 16, and 26—not only had much higher concentrations of SS than the other 29 events, but they also had the inlet loads that represented almost 50 percent of the total load for all events (tables 3–5 and 3–8). Given that the results for the two methods of calculating efficiencies were different at the HSD site and the same at the SFD site, it appears to be important to use both methods at all sites.

Total Suspended-Solids Reductions in Other Field-Verification Studies of the Hydrodynamic-Settling and Stormwater-Filtration Devices

Results from other field-verification studies of the HSD and the SFD might help determine how well the results from this study will apply to other sites. Results from this study indicate that concentrations of SS, TSS, and PSD affect the level of control for a device (figs. 11, 12, 17, and 19). WisDOT would like to know if these study results are transferable from the Riverwalk sites to other sites in Wisconsin. A literature review of previous HSD and SFD studies may be useful in determining the transferability of the results from this study. System performance in this review is described for concentration of TSS only.

Hydrodynamic-Settling Device

Field-verification studies of the HSD have been done in the States of Maine, New Jersey, New York, Connecticut, and Washington (National Cooperative Highway Research Program, 2006). Flow accuracies in the Maine study make it difficult to draw conclusions from the data (Winkler and Guswa, 2002). Data from only five runoff events were collected during the New Jersey study (Greenway, 2001), making it difficult to compare results with those from the HSD study in Milwaukee.

There were 58 runoff events sampled for the study in Connecticut (Clausen and others, 2002). An HSD was installed to treat the runoff from a 1.95-acre school parking lot; eighty percent of the parking lot area was impervious. The SOL determined for TSS was 77 percent, much higher than the 25 percent calculated for the HSD study in Milwaukee. There are two differences in these monitoring projects that could affect the SOLs. First, the parking lot was sanded during the winter months; this might be why 22 of the 58 inlet concentrations of TSS were greater than 250 mg/L, with a maximum value of 3,521 mg/L. The maximum concentration of TSS for the HSD study was 494 mg/L, with only four concentrations greater than 250 mg/L. Only 9 percent of the concentrations of TSS for the HSD study in Milwaukee were greater than 250 mg/L as compared to 38 percent for the Connecticut study. When the inlet concentrations of TSS exceed about 250 mg/L, the efficiency of the device improves and is consistently greater than 50 percent (fig. 17). Second, the inlet water-quality sampling was done with a Coshocton Wheel in the Connecticut study. More information is needed on how the Coshocton Wheel data compare with the automatic-sampler data and how the wheel might affect the magnitude of the concentrations. The Coshocton Wheel results may be more comparable to concentrations of SS.

A HSD was installed in the Village of Lake George, N.Y., to treat the runoff from 9.3 acres of mixed land use, which was considered 95-percent impervious (West and others, 2001; Winkler and Guswa, 2002). About 30 percent of the drainage area was roadway. Samples were collected for

13 runoff events at the inlet and the outlet of the HSD. An external bypass was installed with the device; however, bypass data were not recorded, so the efficiency of the device during bypass events was unclear. An 88 percent SOL was calculated for TSS in the New York study, as compared to 25 percent for the HSD study in Milwaukee.

The HSD study in New York had an average inlet concentration of TSS of 802 mg/L, which was higher than the average of 117 mg/L in the Milwaukee study. In the New York study, approximately 70 percent of inlet concentrations of TSS were greater than 250 mg/L, as compared to approximately 9 percent for the HSD study in Milwaukee. About 38 percent of the inlet concentrations of TSS in the New York study were greater than 1,000 mg/L, with a maximum of 2,492 mg/L. It is not clear why the concentrations of TSS were so high in the New York study runoff samples, but the high concentrations would certainly play a role in the high efficiency seen in the New York study.

The field testing of the HSD in Washington State was done on a 28-acre drainage area along State Route 405 in King County (Taylor Associates, 2002). About 66 percent of the drainage area was estimated to be impervious. Inlet and outlet monitoring was done for 11 runoff events from March 2001 to February 2002. This was the only HSD study reviewed where an effort was made to measure PSD for the samples. The Washington study efficiency ratio was 20 percent for TSS, much lower than the 42 percent measured in the HSD study in Milwaukee. The range in concentrations of TSS in the Washington study was 30 to 580 mg/L, with an average of 190 mg/L, and the range for the HSD study in Milwaukee was 29 to 494 mg/L, with an average of 117 mg/L. Only two concentrations of TSS were greater than 250 mg/L for the Washington study.

Stormwater-Filtration Device

Brown (2003) reviewed six field-verification studies of the SFD. Complications were experienced during each study, making a comparison with this study difficult. The complications included the use of a different filter media, such as leaf compost, and monitoring of less than six events. The findings of this study could be compared with two more recent field-verification studies conducted by the California Department of Transportation (2004) and NSF International (U.S. Environmental Protection Agency, 2005a), because these three studies had similar sampling methods and filter media.

An SFD was installed on a California Department of Transportation maintenance station as part of a BMP retrofit pilot program (California Department of Transportation, 2004). The drainage area of 1.5 acres was 100-percent impervious. A design flow of 2.7 ft³/s was used for the SFD. A mixture of perlite and zeolite was used for the filter media. Particle-size distributions and concentrations of SS were not determined for the runoff samples. The average inlet concentration of TSS of 175 mg/L was similar to the 143 mg/L observed for the SFD study. Only the efficiency ratio was calculated for TSS; and at 40 percent, it was lower than the 59 percent calculated for the SFD study. Without particle-size

data or the concentrations of TSS for each event, it is difficult to speculate as to why the efficiency ratio was higher for the SFD study in Milwaukee.

The field-verification monitoring for the SFD in Griffin, Ga., was part of the USEPA's ETV program (U.S. Environmental Protection Agency, 2005a), as was the SFD study in Milwaukee; hence, both studies used the same monitoring protocols. Instead of freeway, the 0.7-acre drainage area for the SFD study in Georgia was a mixture of parking lot, roadway, and rooftop, with an imperviousness of 85 percent. Perlite was used as a filter media instead of a mixture of zeolite, perlite, and granulated activated carbon. The SFD study in Georgia had a 50-percent SOL for TSS, the same as the SFD study in Milwaukee (table 15). Similar inlet concentrations of TSS observed might be partly responsible for the agreement in the SOLs. The average inlet concentration of TSS for the Georgia study was 165 mg/L, and the average for the Milwaukee study was 143 mg/L. The range in inlet concentrations of TSS of 90 to 410 mg/L for the Georgia study was similar to the range of 22 to 778 mg/L observed for the Milwaukee study (tables 3–5).

Unlike the large differences found among the SOLs measured for TSS and SS in the Milwaukee study, the SOL for both TSS and SS was 50 percent for the Georgia study. This may be explained by the large differences in the PSDs at the two sites. On average, only about 10 percent of the particles in the runoff samples from the Georgia study were in the sand fraction, whereas the average percentage of sand in samples from the SFD in this study was 71 percent.

Mass Balance of Sediment Retained in the Devices

Checking the accuracy of the measured loads at the inlet and outlet of a device requires weighing the material that is retained in the treatment chambers. The weight of the sediment retained in the devices should be reasonably close to the calculated reduction in SS loads. To complete the mass-balance calculation, the SS loads needed to be computed for all events. Ideally, there would be data for concentration of SS for every event during the testing period. Unfortunately, because of the monitoring challenges, there were many events without these data. The HSD and SFD had 59 and 63 unsampled events, respectively. The importance of these unmeasured concentrations was diminished by the fact that precipitation depth for more than one-half of the unsampled events was less than 0.2 in. The goal was to find a method that calculated a reasonable estimate to the measured events and apply that method to unmeasured events, not to match the known sediment retained at the bottom of the devices.

The challenge was to find a method to estimate the inlet and outlet concentrations of SS for the unsampled events. The approach starts with trying to match the concentrations for the sampled events. The SOL for SS using the sampled and unsampled events could then be compared to the weight of the sediment removed from the bottom of the treatment devices.

Table 16. Comparison of suspended-sediment loads estimated with average concentrations and measured suspended-sediment loads for the same runoff events.

[lb, pounds; HSD, hydrodynamic-settling device; SFD, stormwater-filtration device]

Location	Number of sampled events	Measured load (lb)	Estimated load (lb)	Percent difference
HSD				
Inlet	42	182	201	10
Outlet	42	92	77	−16
SFD				
Inlet	32	368	434	16
Outlet	32	40	54	32

Multiple-linear-regression analysis was applied to the runoff events with concentrations of SS in an attempt to estimate SS loads for events with no water-quality data. The regression analysis used flow and precipitation as predictors of concentrations of SS, because flow and precipitation data were available for all events with unmeasured concentrations. The list of independent variables included peak flow; average precipitation intensity; peak 5-, 10-, 15-, 30- and 60-minute precipitation intensity erosivity index; precipitation depth; and antecedent dry days. Similar analysis was completed on the log-transformed concentrations of SS. The regression analysis produced unsatisfactory results in predicting concentrations of SS for the measured events.

The best predictor of the measured data for concentrations of SS proved to be the average concentration of SS measured at the inlet and outlet of each device. The total SS load was determined by multiplying the average concentrations of SS by the measured volumes compared with total measured SS load for the same runoff events (table 16). It was decided that the average concentrations of SS were the best way to determine the SS loads for the unmeasured events.

The average concentration of SS was not used to estimate the outlet load of the HSD for five unmeasured events with peak flows exceeding the design flow of the HSD. Results from four monitored events show the SS efficiency ratios for individual events tend to be negative when the peak flow exceeds the design flow (fig. 16). The average ratio of outlet load to inlet load was about 1.3 for these four monitored events. This ratio was multiplied by the inlet load to estimate the outlet load for the five unmeasured events with peak flows exceeding the design flow. This approach of adjusting outlet loads was not applied to the SFD, because most of the monitored events with peak flows exceeding the design peak flows did not result in negative efficiency ratios (fig. 20). Only event numbers 22 and 24 had negative efficiency ratios, and both of these events had low concentrations of SS relative to the other monitored events.

At the end of the monitoring period, both devices were cleaned out by hand, removing all possible sediment. Standing water was decanted to a level of 0.5 ft above the deposited sediment. Samples of the decanted water were collected at numerous water levels and analyzed for concentrations of SS

and TSS. Sediment removed from each device was collected, then dried and weighed. Subsamples were sent to the USGS Iowa Sediment Laboratory to define the percentage of sediment, by mass, with diameters less than 2,000, 1,000, 500, 250, 125, 62, 31, 16, 8, 4, and 2 μm.

Hydrodynamic-Settling Device

The HSD was cleaned out on September 24, 2004 (figs. 21 and 22). Sediment was removed from the inlet pipe 4-ft upstream from the HSD, the 3-ft-diameter swirl chamber, and the flow-and-oil-control chamber. The dry weight of the sediment at each location was 8 lb for the inlet pipe, 106 lb for the swirl chamber, and 15 lb for the flow-and-oil-control chamber. The total weight from all the locations was 129 lb; most of the sediment retained in the HSD was found in the swirl chamber. The amount of sediment found in the HSD was about the same as predicted by the monitoring data (table 17). Although about one-half of the SS loads had to be estimated, the similarity in the measured and retained loads gave credibility to the monitoring methods.

About 90 percent of the sediment removed from the bottom of the device was in the sand fraction (table 18). The opposite was observed for the inlet water, where 80 percent of the particles were in the silt fraction or smaller (table 6). The difference between the particle sizes in the inlet runoff and the sediment retained by the device clearly shows the device preferentially traps the larger particles and may scour some of the silt particles in subsequent runoff events. A small amount of sediment retained in the inlet pipe and the flow-and-oil-control chamber had a PSD similar to that found in the swirl chamber.

Stormwater-Filtration Device

The SFD was cleaned out on January 24, 2004 (fig. 23). All of the sediment was removed from the inlet bay and the cartridge bay. To determine the amount of material retained by the filters, material from five of the nine cartridges was dried and weighed. The average weight of the cartridges before the study began was subtracted from the total weight at the end of the study. Total sediment retained from the inlet bay was 289 lbs; from the cartridge bay, 145 lbs; and from the filters, 204 lbs; for a total of 638 lbs. The SS load reduction calculated for the SFD using the measured and estimated loads was very close to the amount predicted by weighing the amount of sediment retained in the device's treatment chambers (table 17).

Sediment removed from the SFD inlet and cartridge bays contained particles that were mostly in the sand fraction (table 19). On average, the sediment in the inlet bay was 89-percent sand, and the percent sand in the cartridge bay was 84 percent. The percentage of sand in the filter cartridges was (about 84 percent), nearly the same as in the cartridge bay. A slightly higher percentage of fine particles were trapped in the cartridge bay and filter cartridges than in the inlet bay. Mostly sand was found in the SFD; similarly, the average composition of inlet-water particles was about 71 percent sand (table 8).

Figure 21.   Cleanout of the settling chamber for the hydrodynamic-settling device.

Figure 22.   Cleanout of the 8-inch inlet pipe for the hydrodynamic-settling device.

Table 17. Comparison of the sediment retained, the calculated sum of loads, and the estimated load from the hydrodynamic-settling and stormwater-filtration devices.

[lb, pound; HSD, hydrodynamic-settling device; SFD, stormwater-filtration device]

Type of retained sediment load	HSD loads (lb)	SFD loads (lb)
Calculated sum of loads	90	317
Estimated loads retained	58	334
Total of measured and estimated loads retained	148	651
Amount sediment retained from devices	129	638
Difference between monitored loads and amount of sediment retained from in the devices	19	13
In percent (%)	15%	−2%

Table 18. Particle-size distribution for the sediment samples collected from the hydrodynamic-settling device.

[μm, micrometer; % percent by mass; <, less than; —, not applicable]

Particle size (μm)	Inlet pipe (%)	Swirl chamber subsample 1 (%)	Swirl chamber subsample 2 (%)	Flow-and-oil-control chamber (%)	Median[1] (%)	Average[1] (%)
<8,000	—	—	—	—	—	—
<4,000	88	95	89	87	89	90
<2,000	83	92	84	82	84	85
<1,000	76	87	69	72	74	76
<500	62	74	58	54	60	62
<250	39	40	41	37	40	39
<125	18	16	20	21	19	19
<63	11	7.2	11	14	11	11
<31	6	5.4	8.3	9.8	7	7
<16	4.6	3.8	6.3	6.9	5	5
<8	3.5	2.6	4.3	4.4	4	4
<4	2.8	2.3	3.1	3.4	3	3
<2	2.0	1.6	2.2	2.5	2	2

[1]Statistic combines inlet pipe, swirl chambers 1 and 2, and flow-and-oil-control chamber.

Figure 23. Cleanout of the settling bay and cartridge chamber for the stormwater-filtration device (photograph shows sediment and debris before cleanout).

Table 19. Particle-size distribution for the sediment samples collected from the stormwater-filtration device.

[μm, micrometer; % percent by mass; <, less than; —, not applicable]

Particle Size (μm)	Inlet bay, subsample 1 (%)	Inlet bay, subsample 2 (%)	Inlet bay, subsample 3 (%)	Inlet bay, subsample 4 (%)	Average, inlet bay (%)	Cartridges bay, subsample 1 (%)	Cartridges bay, subsample 2 (%)	Average, cartridges bay (%)
<8,000	—	—	100	96	98	—	—	—
<4,000	92	95	90	92	92	97	95	96
<2,000	83	89	83	87	85	90	89	89
<1,000	69	81	74	75	75	79	75	77
<500	44	65	54	52	54	78	51	65
<250	24	41	31	27	30	32	37	34
<125	13	21	15	16	16	15	28	21
<63	9	14	9	11	11	10	21	16
<31	6.1	8.3	5.9	7.8	7	6.4	16	11
<16	3.8	5.2	4	5.2	5	4.2	11	7
<8	2	3.2	2.3	3.7	3	2.3	7.6	5
<4	1.7	2.6	2	3.1	2	2	6.6	4
<2	1.2	1.9	1.6	2.5	2	1.5	5.3	3

Summary and Conclusions

As part of their efforts to improve the quality of highway runoff, the Wisconsin Department of Transportation (WisDOT) has worked in cooperation with the U.S. Geological Survey, Wisconsin Department of Natural Resources (WDNR), the City of Milwaukee, the Milwaukee Third Ward, Milwaukee County, and the U.S. Environmental Protection Agency Environmental Technology Verification Program to verify the treatment efficiencies of two prefabricated storm-water-treatment devices. The two devices, referred to as the "Riverwalk sites", were installed in December 2001 to treat runoff from a freeway in a high-density urban part of Milwaukee, Wisconsin. Runoff events from June 2002 through October 2004 were monitored for flow and water quality at the inlet and outlet of each device.

One treatment device is categorized as a hydrodynamic-settling device (HSD), which removes pollutants by sedimentation and flotation. The other treatment device is categorized as a stormwater-filtration device (SFD), which removes pollutants by filtration and sedimentation. Filtration is considered the primary method of treatment, with sedimentation of larger particles in the pre-treatment chamber and cartridge filter bay.

Water-quality samples were collected at both the inlet and the outlet of the SFD for 33 runoff events, and 45 for the HSD. For precipitation with depths of 0.2 in. or greater, the percentage of runoff events sampled during the monitoring period was about 70 and 60 percent for the SFD and HSD, respectively. Except for a moderate deviation for precipitation depths between 0.65 and 0.9 in., the distribution of the sampled events was similar to the long-term distribution. Bypassing the system was not possible for the HSD, so all sampled water entered and exited the system. Only a few minutes of bypassing was observed for two events at the SFD site.

Treatment efficiencies of the devices were calculated by means of summation of loads (SOL) and the efficiency-ratio methods. Concentrations and loads of constituents that decreased by passing through the HSD include total suspended solids (TSS), suspended sediment (SS), total phosphorus (TP), total copper (TCu), and total zinc (TZn). The efficiency ratios for these constituents were 42, 57, 17, 33, and 23 percent, respectively. The SOLs for these constituents were 25, 49, 10, 27, and 16 percent, respectively. Concentrations and loads increased at the outlet for chloride (Cl), total dissolved solids (TDS), and dissolved zinc (DZn). The efficiency ratios for these constituents were –347, –177, and –20 percent, respectively. Sand-size particles, rather than silt-size particles, could account for a larger portion of the efficiency ratio and SOLs for SS. The average percentage of sand in the inlet samples was about 30 percent, therefore some of the silt-size particles needed to be captured in the HSD to achieve the SS reduction. Four constituents—dissolved phosphorus (DP), chemical oxygen demand (COD), total polycyclic aromatic hydrocarbon (PAH), and dissolved copper (DCu)—are not included in the list of computed loads, because the difference between the inlet and outlet concentrations for each were not significantly different or too few samples were collected.

Concentrations of constituents and total loads that were decreased by the SFD include TSS, SS, TP, DCu, TCu, DZn, TZn, and COD. The efficiency ratios for these constituents were 59, 90, 40, 21, 66, 23, 66, and 18, respectively. The SOLs for these constituents were 50, 89, 37, 19, 60, 20, 65, and 21, respectively. With the percentage of sand in the inlet runoff averaging 71 percent, the SOLs and efficiency ratios for SS could be more a function of the sand particles than the silt particles, but some of the silt-size particles were retained in the SFD. Similar to the HSD, the average efficiency ratios and SOLs for TDS and Cl were negative. Road-salt brine in both devices appeared to increase the effluent concentrations of both Cl and TDS.

The efficiency of both devices appeared to be consistently higher when concentrations of SS and TSS were greater than 200 mg/L. One possible explanation for the higher efficiencies might be the presence of large particles in samples with higher concentrations of TSS and SS. The larger particles are readily removed by both devices. In contrast, the efficiencies for TSS tend to be negative when the peak flow of an event exceeds the design peak flow for both devices. For the HSD, most of the SS efficiencies also are negative, but only 2 of the 12 events that exceeded the design flow have negative SS efficiencies for the SFD.

The sediment retained inside both devices was removed, weighed, and analyzed for PSD. The concentration of SS was used to predict the amount of sediment that should be retained in the bottom of each device. The amount of sediment predicted to be trapped in the HSD and SFD was about the same as what was removed from the bottom of each device. Most of the sediment retained in both devices was sand- size or larger. The percentage of sand in the swirl chamber for the HSD was 90 percent and in the SFD-filter cartridges was 84 percent. Although the average percentage of sand-size particles at the HSD inlet was only about 30 percent, the device retained mostly sand-size particles. The high percentage of sand observed in the sediment removed from the bottom of the SFD was reflected in the high percentage of sand measured in the inlet water. The cartridge filters did, however, trap some particles ranging in size from 8 to 16 μm.

The WisDOT and the WDNR have an understanding that the WisDOT is to reduce TSS loads in stormwater runoff, and this project provides data on the amount of TSS that might be removed by the HSD and the SFD. The SOL of 25 percent for the HSD and 50 percent for the SFD should approximate the treatment efficiencies expected for TSS at other sections of Wisconsin freeways. Sizing of the devices at other sites would need to reflect careful analysis of the potential peak flows because the devices are not as efficient when the design peak flows are exceeded. The SOLs for TSS are probably best applied to urban freeways instead of rural freeways where the concentrations of TSS and PSD might be different.

The efficiencies for individual events can change with increasing concentration of TSS, but the concentrations of

TSS observed at the Milwaukee Riverwalk sites support the use of the SOLs at other urban freeways in Wisconsin. The average concentrations of TSS measured at the HSD and SFD sites falls within the range of observed concentrations at four other urban-freeway monitoring sites in Wisconsin. The range is small, with the average concentrations of TSS ranging from 106 to 197 mg/L. Based on average concentrations of TSS collected at all six freeway sites, the average concentrations of TSS should not vary enough among urban-freeway sites to appreciably alter the SOL expected for the HSD and the SFD.

Although the sand/silt split data collected at the HSD site compares favorably with the sand/silt split data collected at two other sections of freeway in Milwaukee, the TSS at the SFD site consisted of a relatively high percentage of sand. The average percent of sand-size particles in the runoff at the HSD site was 30 percent, whereas runoff at the SFD site was 71-percent sand. Based on the results from another study of an SFD, the SOL for TSS still would be about 50 percent even if the percent sand was as little as 10 percent. The other study of an SFD in (Griffin, Georgia) of measured a 40-percent reduction in silt-size particles, which might keep the SOL for TSS near 50 percent with just a small amount of sand in the runoff.

Acknowledgments

The authors thank the following people, without whom this project could not have been completed. The staff at the WisDOT for project collaboration and collecting samples throughout the project. James Bachhuber of Earth Tech, Inc., for his assistance with writing the monitoring protocol and maintenance of the devices. David Owens of the USGS, Middleton, Wis., for his instrumentation and programming expertise; and Peter Hughes of the USGS, Middleton, Wis., for his ingenuity in calibrating velocity meters.

References Cited

American Public Health Association and others, 1989, Standard methods for the examination of water and wastewater (17th ed.): Washington, D.C., American Public Health Association [variously paged].

Bachhuber, J., Corsi, S., and Bannerman, R., 2001, Test plan for the verification of Arkal Filtration Systems, Inc.—Pressurized stormwater filtration system, St. Mary's Hospital, Green Bay, Wis.: U.S. Environmental Protection Agency, Office of Research and Development [variously paged].

Bannerman, R.T., Baun, K., Bohn, M., Hughes, P.E., and Graczyk, D.J., 1983, Evaluation of urban nonpoint source pollution management in Milwaukee County, Wisconsin, Volume I for U.S. Environmental Protection Agency, Region V: Wisconsin Department of Natural Resources Publication PB 84–114164 [variously paged].

Bannerman, R.T., Dodds, R.B., Owens, D.W., and Hughes, P.E., 1992, Source of pollutants in Wisconsin stormwater—Volume II for U.S. Environmental Protection Agency, Region V: Wisconsin Department of Natural Resources Grant number C9995007–01 [variously paged].

Brown, Angela, 2003, Development of a BMP Evaluation Methodology for Highway Applications: Corvallis, Oreg., Oregon State University, Master of Science Project Report, Dept. of Civil, Construction, and Environmental Engineering, [variously paged].

Burton, G.A., Jr., and Pitt R.E., 2002, Stormwater effects handbook—A toolbox for watershed managers, scientists, and engineers: Boca Raton, Fla., Lewis Publishers, 929 p.

California Department of Transportation, 2004, BMP Retrofit Pilot Program—Final Report: Division of Environmental Analysis, Report ID CTSW–RT–01–050.

Clausen, J.C., Belanger, P., Board, S., Dietz, M., Phillips, R., and Sonstrom, R., 2002, Stormwater Treatment Devices Section 319 Project: Storrs, Conn., University of Connecticut, Department of Natural Resources Management and Engineering, Project 99–07.

Dong, Allen, Chesters, Gordon, and Simsiman, G.V., 1979, Dispersibility of soils and elemental composition of soils, sediments, and dust and dirt from the Menomonee River Watershed: U.S. Environmental Protection Agency Report EPA–905/4–79–029–F, 56 p.

Driscoll, E.D., Shelley, P.E., and Strecker, E.W., 1990, Pollutant loadings and impacts from highway stormwater runoff, Volume I—Design procedure: Federal Highway Administration Final Report FHWA–RD–88–006, 61 p.

Gray, J.R., Glysson, D.G., Turcois, L.M., and Schwarz, G.E., 2000, Comparability of suspended-sediment concentrations and total suspended solids data: U.S. Geological Survey Water-Resources Investigations Report 00–4191, 14 p.

Greenway, R.A., 2001, Stormwater Treatment Demonstration Project—Oil water/swirl separator followed by a sand filter: RTP Environmental Associates, Inc., prepared for Harding Township, N.J., Environmental Commission and the New Jersey Department of Environmental Protection Paper WM–668.

Gupta, M.K., Agnew, R.W., and Kobriger, N.P., 1981, Constituents of highway runoff, volume I, State-of-the-art report: U.S. Federal Highway Administration Report FHWA/RD–81/042, 111 p.

Guy, H.P., 1977, Laboratory theory and methods for sediment analysis: U.S. Geological Survey Techniques of Water-Resources Investigations, book 5, chap. C1, 58 p.

Helsel, D.R., 2004, Nondetects and data analysis—Statistics for censored environmental data: Wiley–Interscience, 268 p.

Helsel, D.R., and Hirsch, R.M., 1992, Statistical methods in water resources: New York, Elsevier, 522 p.

Horwatich, J.A., Corsi, R.S., and Bannerman, R.T., 2004, Effectiveness of a pressurized stormwater filtration system in Green Bay, Wisconsin—A study for the Environmental Technology Verification Program of the U.S. Environmental Protection Agency: U.S. Geological Survey Scientific Investigations Report 2004–5222, p. 19.

Knott, J.M., Glysson, G.D., Malo, B.A., and Schroder, L.J., 1993, Quality assurance plan for the collection and processing of sediment data: U.S. Geological Survey Open-File Report 92–499, 18 p.

Kopp, J.F., and McKee, G.D., 1979, Methods for chemical analysis of water and waste (3d ed.): U.S. Environmental Protection Agency, Environmental Monitoring and Support Laboratory, EPA–600/4–79–020 [variously paged].

Mahler B.J., Van Meter, P.C., Bashara T.J., Wilson J.T., and Johns D. A., 2005, Parking lot sealcoat—An unrecognized source of urban polycyclic aromatic hydrocarbons: Environmental Science & Technology, v. 39, no. 15, p. 5560–5566.

National Cooperative Highway Research Program, 2006, Evaluation of best management practices for highway runoff control: Washington, D.C., Transportation Research Board, NCHRP Report 565, 111 p., 2 app.

National Oceanic and Atmospheric Administration, 2007a, National Climatic Data Center, Milwaukee Mitchell Airport Weather Service Office (WSO) rainfall records, 1949–1992 and 2002–2004.

National Oceanic and Atmospheric Administration, 2007b, National Climatic Data Center, Mt. Mary College, Milwaukee, Wis., rainfall records, 2002–2004.

Pitt, R., 1987, Small storm urban flow and particulate wash-off contributions to storm-sewer outfall discharges: University of Wisconsin–Madison, Department of Civil and Environmental Engineering, 513 p.

Rantz S.E., and others, 1982, Measurement and computation of streamflow—v. 2, Computation of discharge: U.S. Geological Survey Water-Supply Paper 2175, p. 285–631.

Rickert, D.A., 1997, Comparison of the suspended-sediment splitting capabilities of the churn and cone splitters: U.S. Geological Survey Office of Water Quality Technical Memorandum No. 97.06, accessed June 1, 2009, at *http://water.usgs.gov/admin/memo/QW/qw97.06.html*

Selbig, W.R., and Bannerman, R.T., 2007, Evaluation of street sweeping as a stormwater-quality-management tool in three residential basins in Madison, Wisconsin: U.S. Geological Survey Scientific Investigations Report 2007–5156, 104 p.

Taylor Associates, 2002, SR 405 Vortechs water quality monitoring report: Olympia, Wash., Report to the Washington State Department of Transportation.

Thomson, N.R., McBean, E.A., Snodgrass, W., and Monstrenko, I.B., 1997, Highway stormwater runoff quality—Development of surrogate parameter relationships: Water, Air, and Soil Pollution, v. 94, nos. 3–4, p. 307–347.

U.S. Environmental Protection Agency, 1983, Results of the Nationwide Urban Runoff Program, Volume 1—Final report: Washington, D.C., Water Planning Division, available from the National Technical Information Service as PB84–185552 [variously paged].

U.S. Environmental Protection Agency, 1999, Preliminary data summary of urban storm water best management practices: U.S. Environmental Protection Agency EPA–821–R–99–012 [variously paged].

U.S. Environmental Protection Agency, 2000, Storm water phase II final rule—An overview: U.S. Environmental Protection Agency EPA 833–F–00–001, Fact Sheet 1.0, 4 p.

U.S. Environmental Protection Agency, 2002, ETV verification protocol, stormwater source area treatment technologies, EPA/NSF wet-weather flow technologies pilot: v. 4.1 [variously paged].

U.S. Environmental Protection Agency, 2004, Environmental technology verification report—Stormwater source area treatment device—The stormwater management Storm-Filter® using ZPG filter media: 04/17/WQPC–WWF, EPA/600/R–04/125, 65 p., accessed June 1, 2009, at *http://www.nsf.org/business/water_quality_protection_center/pdf/SMI_Riverwalk_Verification_Report_Final.pdf*

U.S. Environmental Protection Agency, 2005a, Environmental technology verification report—Stormwater source area treatment device—The stormwater management Storm-Filter® using perlite filter media: 05/23/WQPC–WWF, EPA 600/R–05/137, 56 p., accessed June 1, 2009, at *http://www.nsf.org/business/water_quality_protection_center/pdf/StormFilter_Griffin_Report.pdf*

U.S. Environmental Protection Agency, 2005b, Environmental technology verification report—Stormwater source area treatment device—Vortechnics, Inc., Vortechs® system, model 1000: 05/24/WQPC–WWF, EPA 600/R–05/140, 66 p., accessed June 1, 2009, at *http://www.nsf.org/business/water_quality_protection_center/pdf/Vortechs_Verification_Report.pdf*

Waschbusch, R.J., 1996, Stormwater-runoff data in Madison, Wisconsin, 1993–94: U.S. Geological Survey Open-File Report 95–733, 33 p.

Waschbusch, R.J., 1999, Evaluation of the effectiveness of urban stormwater treatment unit in Madison, Wisconsin, 1996–97: U.S. Geological Survey Water-Resources Investigations Report 99–4195, 49 p.

Waschbusch, R.J., 2003, Data and methods of a 1999–2000 street sweeping study on an urban freeway in Milwaukee County, Wisconsin: U.S. Geological Survey Open-File Report 2003–93, 41 p.

West, T.A., Bloomfield, J.A., and Lake, D.W., Jr., 2001, Final report—A study of the effectiveness of a Vortechs stormwater treatment system for removal of total suspended solids and other pollutants in the Marine Village Watershed, Village of Lake George, New York: New York State of Environmental Conservation.

Winkler, E.S., and Guswa, Susan, 2002, Final Technology Assessment Report—Vortechs® Stormwater Treatment System , Vortechnics, Inc., Scarborough, Maine: Amherst, Mass., University of Massachusetts at Amherst, Center for Renewable Energy Efficiency and Renewable Energy, prepared for the Strategic Envirotechnology Partnership, 38 p.

Wisconsin Department of Transportation, 2002, Construction site erosion control and storm water management procedures for department actions: Wisconsin Administrative Code, chap. TRANS 401.03 [variously paged].

Woodworth, M.T., and Connor, B.F., 2003, Results of the U.S. Geological Survey's analytical evaluation program for standard reference samples distributed in March 2003: U.S. Geological Survey Open-File Report 2003–261, 109 p.

Appendix 1.   Previous Studies

Bachhuber, J.A., Horwatich, J.A., Corsi, S.R., and Bannerman, R.T., 2002, Environmental Technology Verification Program: Development of a Protocol for Testing Commercial Stormwater Treatment Devices and Two Case Examples in Wisconsin, Conference proceedings paper for StormCon 2002, Marco Island, FL, August, 2002.

Bannerman, R.T., Baun, K., Bohn, M., Hughes, P.E., and Graczyk, D.J., 1983, Evaluation of urban nonpoint source pollution management in Milwaukee County, Wisconsin, Volume 1 for U.S. Environmental Protection Agency, Region V: Wisconsin Department of Natural Resources Publication PB 84–114164 [variously paged]

Bannerman, R.T., Dodds, R.B., Owens, D.W., Hughes, P.E., 1992, Source of pollutants in Wisconsin Stormwater: 1 for U.S. Environmental Protection Agency Region V: Wisconsin Department of Natural Resources Grant number C9995007–01 [variously paged].

Bannerman, R.T., Owens, D.W., Dodds, R.B., and Hornewer, N.J., 1993, Sources of pollutants in Wisconsin stormwater: Water Science Technology, v. 28, no. 3–5, p. 241–259.

Bannerman, R.T., Legg, A.D., and Greb, S.R., 1996, Quality of Wisconsin stormwater 1989—94: U.S. Geological Survey Open-File Report 96–458, 26 p.

Corsi, S.R., Graczyk, D.J., Owens, D.W., and Bannerman, R.T., 1997, Unit-area loads of suspended sediment, suspended solids, and total phosphorus from small watersheds in Wisconsin: U.S. Geological Survey Fact Sheet 195–97, 4 p.

Corsi, S.R., Greb, S.R., Bannerman, R.T., and Pitt, R.E., 1999, Evaluation of the multi-chambered treatment train, a retrofit water-quality management practice: U.S. Geological Survey Open-File Report 99–270, 24 p.

Corsi, S.R., Walker, J.F., Wang, L., Horwatich, J.A., and Bannerman, R.T., 2005, Effects of best-management practices in Otter Creek in the Sheboygan River Priority Watershed, Wisconsin, 1990–2002: U.S. Geolocial Survey Scientific Investigations Report 2005–5009.

Graczyk, D.G., Hunt, R.J., Greb, S.R., Buchwald, C.A., and Krohelski, J.T., 2003, Hydrology, nutrient concentrations, and nutrient yields in nearshore areas of four lakes in northern Wisconsin, 1999–2001: U.S. Geological Survey Water-Resources Investigations Report 03–4144, 64 p.

Graczyk, D.J., Walker, J.F., Horwatich, J.A., and Bannerman, R.T., 2003, Effects of best-management practices in the Black Earth Creek priority watershed, Wisconsin, 1984–98: U.S. Geological Survey Water-Resources Investigations Report 03–4163, 24 p.

Greb, S.R., Bannerman, R.T., Corsi, S.R., Pitt, R.E., 2000, Evaluation of the Multichambered Treatment Train, a Retrofit Water-Quality Management Practice, Water Environmental Research, vol. 72, no. 2, 207–216.

Horwatich, J.A., and Bannerman, R.T., 2010, Parking lot runoff quality and treatment efficiency of a stormwater-filtration device, Madison, Wisconsin, 2005–07: U.S. Geological Survey Scientific Investigations Report 2009–5196, p 50.

Horwatich, J.A., Corsi, R.S., Bannerman, R.T., 2004, Effectiveness of a pressurized stormwater filtration system in Green Bay, Wisconsin—A study for the Environmental Technology Verification Program of the U.S. Environmental Protection Agency: U.S. Geological Survey Scientific Investigations Report 2004–5222, p 19.

House, L.B., Waschbusch, R.J., Hughes, P.E., 1993, Water quality of an urban wet detention pond in Madison Wisconsin, 1987–88: U.S. Geological Survey Open-File Report 93–172, 57 p.

Legg, A.D., Bannerman, R.T., and Panuska, J., 1996, Variation in the relation of rainfall to runoff from residential lawns in Madison, Wisconsin, July and August 1995: U.S. Geological Survey Scientific Investigations Report 96–4196, 11 p.

Martinelli, T.J., Waschbusch, R.J., Bannerman, R.T., and Wisner, A., 2002, Pollutant loading to stormwater runoff from highways: impact of a freeway sweeping program: WI Department of Transportation Final Report WI–11–01, 94 p.

Owens, D.O., Jopke, P., Hall, D.W., Balousek, J., and Roa, A., 2000, Soil erosion from two small construction sites, Dane County, Wisconsin: U.S. Geological Survey Fact Sheet FS–109–00, 4 p.

Robertson, D.M., Graczyk, D.J., Garrison, P.J., Wang, L., LaLiberte, G., and Bannerman, R.T., 2006, Nutrient Concentrations and Their Relations to the Biotic Integrity of Wadeable Streams in Wisconsin: U.S. Geological Survey Professional Paper 1722, 156 p.

Selbig, W.R., Bannerman, R.T., and Bowman, G. 2006. Use of Wet Sieving to Improve the Accuracy of Sediment and Sediment-Associated Constituent Concentrations in Whole-Water Samples. Proceedings of the 8th Federal Interagency Sedimentation and Hydrologic Modeling Conference, Reno, NV, April, 2006.

Selbig, W.R. and Bannerman, R.T., 2007, Evaluation of street sweeping as a stormwater-quality management tool in three residential basins in Madison, Wisconsin, U.S. Geological Survey Scientific Investigations Report 2007–5, 103 p.

Selbig, W.R., Bannerman, R.T., and Bowman, G., 2007, Improving the Accuracy of Sediment-Associated Constituent Concentrations in Whole Storm Water Samples by Wet Sieving, Journal of Environmental Quality, vol. 36, no. 1, 7 p.

Selbig, W.R., and Bannerman, R.T., 2008, A Comparison of Runoff Quantity and Quality from Two Small Basins Undergoing Implementation of Conventional and Low-Impact-Development (LID) Strategies: Cross Plains, Wisconsin, Water Years 1999–2005: U.S. Geological Survey Scientific Investigations Report 2008–5008, 66 p.

Steuer, J.J., Selbig, W.R., Hornewer, N.J., and Prey, J., 1997, Sources of contamination in an urban basin in Marquette, Michigan, and an analysis of concentrations, loads, and data quality: U.S. Geological Survey Water-Resources Investigations Report 97–4242, 25 p.

Steuer, J.J., Selbig, W.R., and Hornewer, N.J., 1996, Contaminant concentration in stormwater from eight Lake Superior basin cities, 1993–94: U.S. Geological Survey Open-File Report 96–122, 16 p.

Stewart, J.S., Wang, L., Lyons, J., Wierl, J.A., and Bannerman, R.T., 2001, Influences of watershed, riparian corridors, and reach-scale characteristics on aquatic biota in agricultural watersheds. Journal of the American Water Resources Association 37: 1475–1487.

Stuntebeck, T.D., 1995, Evaluating barnyard best management practices in Wisconsin using upstream-downstream monitoring: U.S. Geological Survey Fact Sheet 221–95.

Stuntebeck, T.D., and Bannerman, R.T., 1998, Effectiveness of Barnyard Best Management Practices in Wisconsin: U.S. Geological Survey Fact Sheet 051–98.

Stuntebeck, T.D., Komiskey, M.J., Owens, D.W., and Hall, D.W., 2008, Methods of Data Collection, Sample Processing, and Data Analysis for Edge-of-Field, Streamgaging, Subsurface-Tile, and Meteorological Stations at Discovery Farms and Pioneer Farm in Wisconsin, 2001–7: U.S. Geological Survey Open-File Report 2008–1015, 60 p.

U.S. Environmental Protection Agency, 1982, Results of the Nationwide Urban Runoff Program—Volume II, Appendices: U.S. Environmental Protection Agency [variously paginated].

U.S. Environmental Protection Agency, 1983, Results of the Nationwide Urban Runoff Program, Volume 1, Final Report: Water Planning Division, Washington, D.C. [variously paged].

U.S. Environmental Protection Agency, 2004a, Environmental Technology Verification Report Stormwater source area treatment device-Arkal Pressurized Stormwater Filtration System: U.S. Environmental Protection Agency EPA/600/R–04/084 [variously paged].

U.S. Environmental Protection Agency, July 2004b, Environmental Technology Verification Report—Stormwater source area treatment device—The stormwater management StormFilter using ZPG filter media: 04/17/WQPC-WWF, EPA/600/R–04/125, 65 p.

U.S. Environmental Protection Agency, 2005a, Environmental Technology Verification Report— Stormwater management StormFilter using perlite filter media: 05/23/WQPC-WWF, EPA 600/R–05/137, 56 p.

U.S. Environmental Protection Agency, 2005b, Environmental Technology Verification Report—Stormwater source area treatment device—Vortechnics, Inc., Vortechs system, model 1000: 05/24/WQPC-WWF, EPA 600/R–05/140, 66 p.

Walker, J.F., Graczyk, D.J., Corsi, S.R., Owens, D.W., and Wierl, J.A., 1995, Evaluation of nonpoint-source contamination, Wisconsin; land-use and best-management-practices inventory, selected streamwater-quality data, urban-watershed quality assurance and quality control, constituent loads in rural streams, and snowmelt-runoff analysis, water year 1994: U.S. Geological Survey Open-File Report 95–320, 21 p.

Waschbusch, R.J., 1995, Stormwater-runoff data in Madison, Wisconsin, 1993–94: U.S. Geological Survey Open-File Report 95–733, 33 p.

Waschbusch, R.J., 1999, Evaluation of the effectiveness of urban stormwater treatment unit in Madison, Wisconsin, 1996–97: U.S. Geological Survey Water-Resources Investigations Report 99–4195, 49 p.

Waschbusch, R.J., 2003, Data and Methods of a 1999–2000 street sweeping study on an urban freeway in Milwaukee County, Wisconsin: U.S. Geological Survey Open-File Report 03–93, 41 p.

Waschbusch, R.J., Bannerman, R.T., and Greb, S.R., 1997, Yields of selected constituents in street runoff in Madison Wisconsin, 1994–95: Written Comm. to the Wisconsin Department of Natural Resources.

Waschbusch, R.J., Selbig, W.R, and Bannerman, R.T., 1999, Sources of phosphorus in stormwater and street dirt from two urban residential basins in Madison, Wisconsin, 1994–95: U.S. Geological Survey Water-Resources Investigations Report 99–4021, 47 p.

Weigel, B.M., Emmons, E., Stewart, J.S., and Bannerman, R., 2005. Buffer width and continuity for preserving stream health in agricultural landscapes, Wisconsin Department of Natural Resources Research/Management Findings, Wisconsin Department of Natural Resources, Madison, WI, PUB–SS–756 2005, 4 p.

Wierl [Horwatich], J.A., Giddings, E.M.P., Bannerman, R.T., 1998, Evaluation of a Method for Comparing Phosphorus Loads from Barnyards and Croplands in Otter Creek Watershed, Wisconsin: U.S. Geological Survey Fact Sheet FS 168–98, 4 p.

Wierl, [Horwatich], J.A., Rappold, K.F., and Amerson, F.U., 1996, Summary of the land-use inventory for the nonpoint-source evaluation monitoring watersheds in Wisconsin: U.S. Geological Survey Open-File Report 96–123, 23 p.

Appendix 2.   Hydrodynamic-Settling Device

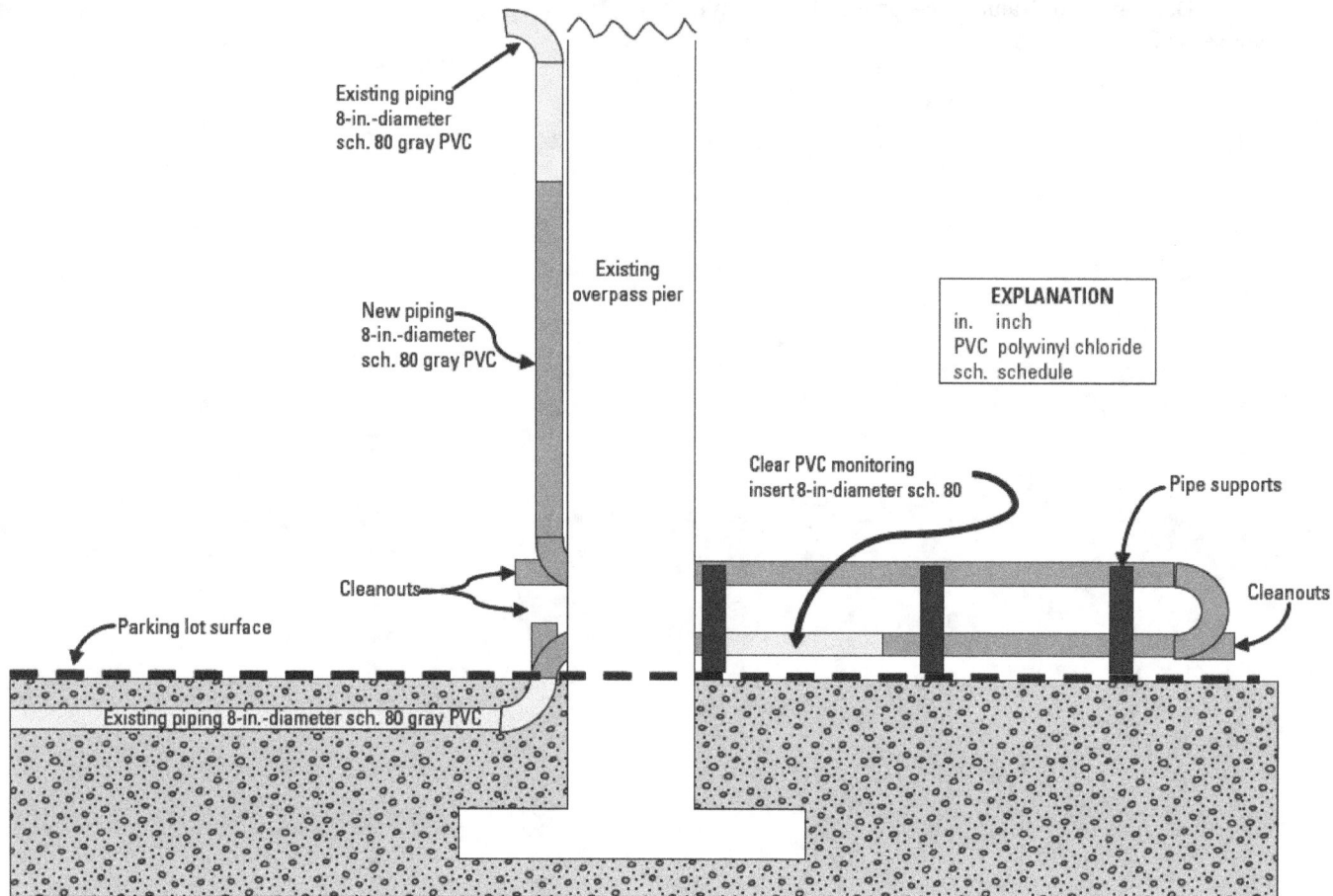

Figure 2–1.   Piping modifications to the hydrodynamic-settling device, Milwaukee, Wis.

Table 2–1.   Rainfall data for monitored events, hydrodynamic-settling device, Milwaukee, Wis.

[mm, month; dd, day; yyyy, year; hh:mm, hour and minutes; in., inch; min, minute; in/h, inch per hour; ft-lb/acre/in/h, foot-pound per acre per inch per hour; ft³, cubic foot; —, not computed for event; ; GMIA, General Mitchell International Airport]

Sampled event number	Start date and time (mm/dd/yyyy hh:mm)	End date and time (mm/dd/ yyyy hh:mm)	Rainfall duration (hh:mm)	Total rainfall (in.)	Max 15-min intensity (in/h)	Max 30-min intensity (in/h)	Erosivity index (hundreds of ft-lb/ acre/in/hr)	Rainfall volume (ft³)	Antecedent dry times (dd hh:mm)	Comments
	04/30/2003 07:54	4/30/2003 08:36	00:42	0.08	0.12	0.12	0.08	73	08 21:44	
1	04/30/2003 13:30	4/30/2003 14:30	01:00	.35	.76	.54	1.7	318	00 04:54	
2	04/30/2003 22:08	05/01/2003 01:38	03:30	1.1	1.0	.88	8.4	989	00 07:38	
	05/01/2003 11:19	05/01/2003 14:11	02:52	.08	.12	.08	.05	73	00 09:41	
3	05/04/2003 21:21	05/05/2003 01:26	04:05	.72	.36	.30	1.8	653	03 07:10	
4	05/05/2003 04:14	05/05/2003 09:05	04:51	.17	.24	.20	.29	154	00 02:48	
	05/07/2003 05:36	05/07/2003 06:35	00:59	.12	.16	.14	.14	109	01 20:31	
	05/07/2003 11:54	05/07/2003 17:15	05:21	.26	.16	.12	.26	236	00 05:19	
5	05/09/2003 00:12	05/09/2003 04:39	04:27	.87	.60	.42	3.1	790	01 06:57	
	05/11/2003 12:39	05/11/2003 19:57	07:18	.16	.08	.08	.11	145	02 08:00	
	05/14/2003 11:39	05/14/2003 12:53	01:14	.05	.12	.06	.03	45	02 15:42	
	05/14/2003 16:49	05/15/2003 01:14	08:25	.23	.12	.10	.19	209	00 03:56	
	05/15/2003 06:01	05/15/2003 08:03	02:02	.03	.04	.04	.01	27	00 04:47	
6	05/20/2003 00:16	05/20/2003 02:41	02:25	.19	.16	.14	.22	172	04 16:13	
7	05/30/2003 18:54	05/30/2003 23:01	04:07	.54	.52	.32	1.5	490	10 16:13	
	05/31/2003 05:11	05/31/2003 05:28	00:17	.13	.48	—	—	118	00 06:10	
8	06/08/2003 03:26	06/08/2003 14:35	11:09	.62	.80	.54	2.1	563	07 21:58	
9	06/27/2003 17:30	06/27/2003 20:08	02:38	.37	.60	.40	1.3	336	19 02:55	
9	06/28/2003 08:29	06/28/2003 10:55	02:26	.20	.36	.22	.39	182	00 12:21	
10	07/04/2003 07:25	07/04/2003 08:57	01:32	.15	.52	.26	.35	136	05 20:30	
10	07/05/2003 04:33	07/05/2003 06:14	01:41	.31	.36	.32	.84	281	00 19:36	
10	07/06/2003 09:30	07/06/2003 10:08	00:38	.07	.20	.12	.07	64	01 03:16	
11	07/06/2003 15:06	07/06/2003 16:19	01:13	0.14	0.36	0.20	0.24	127	00 04:58	

Table 2–1.   Rainfall data for monitored events, hydrodynamic-settling device, Milwaukee, Wis.—Continued

[mm, month; dd, day; yyyy, year; hh:mm, hour and minutes; in., inch; min, minute; in/h, inch per hour; ft-lb/acre/in/h, foot-pound per acre per inch per hour; ft³, cubic foot; —, not computed for event; ; GMIA, General Mitchell International Airport]

Sampled event number	Start date and time (mm/dd/yyyy hh:mm)	End date and time (mm/dd/yyyy hh:mm)	Rainfall duration (hh:mm)	Total rainfall (in.)	Max 15-min intensity (in/h)	Max 30-min intensity (in/h)	Erosivity index (hundreds of ft-lb/ acre/in/hr)	Rainfall volume (ft³)	Antecedent dry times (dd hh:mm)	Comments
	07/06/2003 19:49	07/06/2003 20:02	00:13	.03	—	—	—	27	00 03:30	
	07/07/2003 08:20	07/07/2003 08:49	00:29	.10	.32	—	—	91	00 12:18	
12	07/08/2003 09:49	07/08/2003 13:26	03:37	.33	.24	.20	.56	299	01 01:00	
13	07/09/2003 23:14	07/10/2003 00:43	01:29	.07	.24	.12	.08	64	01 09:48	
14	07/15/2003 02:56	07/15/2003 04:46	01:50	.17	.20	.12	.17	154	05 02:13	
	07/21/2003 09:32	07/21/2003 10:14	00:42	.19	.72	.36	.66	172	06 04:46	
15	07/30/2003 15:14	07/30/2003 19:45	04:31	.19	.64	.34	.73	172	09 05:00	
16	08/01/2003 00:30	08/01/2003 02:54	02:24	.13	.40	.22	.26	118	01 04:45	
16	08/01/2003 06:03	08/01/2003 06:10	00:07	.10	—	—	—	91	00 03:09	
16	08/02/2003 17:38	08/02/2003 17:47	00:09	.09	—	—	—	82	01 11:28	
16	08/03/2003 12:34	08/03/2003 14:21	01:47	.41	.64	.50	1.8	372	00 18:47	
	08/11/2003 22:54	08/11/2003 23:41	00:47	.11	.40	.20	.19	100	08 08:33	
17	08/25/2003 18:49	08/25/2003 19:36	00:47	.30	1.2	.58	1.8	272	13 19:08	
18	09/12/2003 15:32	09/12/2003 19:21	03:49	.30	.24	.22	.56	272	17 19:56	
	09/13/2003 07:30	09/13/2003 10:52	03:22	.16	.16	.12	.16	145	00 12:09	
19	09/14/2003 05:22	09/14/2003 11:57	06:35	.47	1.4	.16	.16	427	00 18:30	
	09/22/2003 02:28	09/22/2003 06:05	03:37	.27	.32	.24	.67	245	07 14:31	
20	09/26/2003 16:11	09/26/2003 19:23	03:12	.15	.16	.14	.18	136	04 10:06	
21	10/03/2003 10:15	10/03/2003 12:23	02:08	.14	.12	.12	.14	127	06 14:52	
22	10/11/2003 21:58	10/12/2003 00:02	02:04	.11	.08	.08	.07	100	08 09:35	
23	10/14/2003 00:17	10/14/2003 03:10	02:53	.27	.20	.16	.36	245	02 00:15	
24	10/14/2003 07:08	10/14/2003 09:49	02:41	0.23	0.24	0.20	0.39	209	00 03:58	Rainfall from GMIA
25	10/24/2003 16:45	10/24/2003 22:16	05:31	.71	.36	.34	2.0	644	10 06:56	

Table 2–1. Rainfall data for monitored events, hydrodynamic-settling device, Milwaukee, Wis.—Continued

[mm, month; dd, day; yyyy, year; hh:mm, hour and minutes; in., inch; min, minute; in/h, inch per hour; ft-lb/acre/in/h, foot-pound per acre per inch per hour; ft³, cubic foot; —, not computed for event; ; GMIA, General Mitchell International Airport]

Sampled event number	Start date and time (mm/dd/yyyy hh:mm)	End date and time (mm/dd/ yyyy hh:mm)	Rainfall duration (hh:mm)	Total rainfall (in.)	Max 15-min intensity (in/h)	Max 30-min intensity (in/h)	Erosivity index (hundreds of ft-lb/ acre/in/hr)	Rainfall volume (ft³)	Antecedent dry times (dd hh:mm)	Comments
	11/01/2003 22:06	11/02/2003 08:05	09:59	.63	.32	.24	1.3	572	07 23:50	
	11/04/2003 16:14	11/04/2003 20:21	04:07	.60	.68	.36	1.4	545	02 08:09	
	11/17/2003 23:10	11/18/2003 12:11	13:01	1.08	.52	.40	3.6	980	13 02:49	
	11/22/2003 17:26	11/22/2003 21:59	04:33	.12	.12	.08	.08	109	04 05:15	
	11/23/2003 05:37	11/23/2003 15:04	09:27	.13	.12	.10	.11	118	00 07:38	
	12/09/2003 12:30	12/10/2003 16:57	04:27	1.9	.32	.24	3.8	1,724	15 21:26	
	12/16/2003 03:29	12/16/2003 04:58	01:29	.11	.16	.12	.11	100	05 10:32	
26	12/28/2003 01:06	12/28/2003 05:34	04:28	.22	.16	.12	.22	200	11 20:08	
	03/01/2004 15:39	03/01/2004 17:10	01:31	.15	.24	.16	.20	136	04 10:05	
	03/04/2004 16:41	03/05/2004 06:22	13:41	1.89	.44	.40	6.4	1,715	02 23:31	
	03/13/2004 23:16	03/14/2004 03:46	04:30	.16	.08	.06	.08	145	08 16:54	
	03/17/2004 11:38	03/17/2004 18:13	06:35	.08	.04	.04	.03	73	03 07:52	
	03/18/2004 09:18	03/18/2004 17:04	07:46	.14	.08	.06	.07	127	00 15:05	
27	03/25/2004 22:59	03/26/2004 03:56	04:57	.85	.48	.36	2.63	771	07 05:55	
28	03/28/2004 15:18	03/28/2004 20:07	04:49	.87	.48	.42	3.12	790	02 11:22	
	03/30/2004 05:21	03/30/2004 12:41	07:20	.13	.08	.06	.07	118	01 09:14	
29	04/17/2004 02:53	04/17/2004 04:11	01:18	.24	.44	.34	.69	218	17 14:12	
30	04/20/2004 16:21	04/21/2004 03:12	10:51	1.41	1.16	.78	9.5	1,280	03 12:10	
	04/24/2004 23:26	04/25/2004 00:21	00:55	.08	.20	.16	.49	73	03 20:14	
	04/25/2004 15:26	04/25/2004 15:33	00:07	.07	—	—	—	64	00 15:05	
	04/30/2004 19:31	04/30/2004 23:09	03:38	0.08	0.08	0.06	0.04	73	05 03:58	
	05/07/2004 19:06	05/07/2004 20:10	01:04	.07	.12	.08	.05	64	06 19:57	
	05/08/2004 22:04	05/09/2004 02:53	04:49	.44	1.16	.66	2.7	399	01 01:54	

Table 2–1.   Rainfall data for monitored events, hydrodynamic-settling device, Milwaukee, Wis.—Continued

[mm, month; dd, day; yyyy, year; hh:mm, hour and minutes; in., inch; min, minute; in/h, inch per hour; ft-lb/acre/in/h, foot-pound per acre per inch per hour; ft³, cubic foot; —, not computed for event; ; GMIA, General Mitchell International Airport]

Sampled event number	Start date and time (mm/dd/yyyy hh:mm)	End date and time (mm/dd/ yyyy hh:mm)	Rainfall duration (hh:mm)	Total rainfall (in.)	Max 15-min intensity (in/h)	Max 30-min intensity (in/h)	Erosivity index (hundreds of ft-lb/ acre/in/hr)	Rainfall volume (ft³)	Antecedent dry times (dd hh:mm)	Comments
	05/10/2004 14:46	05/10/2004 18:27	03:41	1.19	1.20	.72	7.7	1,080	01 11:53	
31	05/12/2004 18:22	05/13/2004 03:27	09:05	.55	.44	.34	1.7	499	01 23:55	
	05/13/2004 17:06	05/13/2004 17:49	00:43	1.79	4.96	3.40	47.0	1,624	00 13:39	
	05/14/2004 03:29	05/14/2004 12:31	09:02	1.36	.60	.50	5.9	1,234	00 09:40	
	05/17/2004 22:05	05/18/2004 02:44	04:39	.57	.64	.34	1.67	517	03 09:34	
32	05/20/2004 16:31	05/20/2004 17:33	01:02	.24	.72	.36	.80	218	02 13:47	
	05/21/2004 08:55	05/21/2004 10:03	01:08	.70	1.64	1.12	7.4	635	00 15:22	
33	05/21/2004 17:34	05/22/2004 08:07	14:33	1.78	1.40	.98	15.6	1,615	00 07:31	
	05/22/2004 19:24	05/23/2004 07:24	12:00	1.15	2.28	1.16	13.3	1,044	00 11:17	
	05/25/2004 03:55	05/25/2004 05:34	01:39	.07	.08	.06	.04	64	01 20:31	
	05/29/2004 08:44	05/29/2004 12:48	04:04	.17	.08	.08	.12	154	04 03:10	
34	05/30/2004 10:48	05/30/2004 12:59	02:11	.68	.56	.48	2.8	617	00 22:00	
	05/30/2004 19:00	05/31/2004 05:56	10:56	.18	.36	.18	.29	163	00 06:01	
	05/31/2004 12:25	05/31/2004 22:46	10:21	.21	.48	.24	.45	191	00 06:29	
	06/08/2004 23:15	06/08/2004 23:20	00:05	.10	—	—	—	91	08 00:29	
	06/09/2004 05:55	06/09/2004 08:42	02:47	.07	.24	.12	.07	64	00 06:35	
35	06/10/2004 03:21	06/11/2004 12:19	08:58	1.72	.52	.40	5.8	1,561	00 18:39	
	06/11/2004 21:50	06/12/2004 03:50	06:00	.23	.24	.20	.39	209	00 09:31	
	06/14/2004 02:12	06/14/2004 02:26	00:14	.11	—	—	—	100	01 22:22	
36	06/14/2004 11:27	06/14/2004 12:26	00:59	0.82	2.96	1.56	13.6	744	00 09:01	
37	06/16/2004 19:37	06/16/2004 20:04	00:27	.10	.32	—	—	91	02 07:11	
	06/17/2004 05:06	06/17/2004 08:19	03:13	.22	.20	.18	.33	200	00 09:02	

Table 2–1. Rainfall data for monitored events, hydrodynamic-settling device, Milwaukee, Wis.—Continued

[mm, month; dd, day; yyyy, year; hh:mm, hour and minutes; in., inch; min, minute; in/h, inch per hour; ft-lb/acre/in/h, foot-pound per acre per inch per hour; ft³, cubic foot; —, not computed for event; ; GMIA, General Mitchell International Airport]

Sampled event number	Start date and time (mm/dd/yyyy hh:mm)	End date and time (mm/dd/yyyy hh:mm)	Rainfall duration (hh:mm)	Total rainfall (in.)	Max 15-min intensity (in/h)	Max 30-min intensity (in/h)	Erosivity index (hundreds of ft-lb/acre/in/hr)	Rainfall volume (ft³)	Antecedent dry times (dd hh:mm)	Comments
	06/21/2004 09:37	06/21/2004 15:41	06:04	.66	.32	.28	1.56	599	04 01:18	
	06/23/2004 12:26	06/23/2004 14:30	02:04	.07	.08	.06	.04	64	01 20:45	
38	06/24/2004 09:12	06/24/2004 12:49	03:37	.23	.20	.16	.31	209	00 18:42	
39	06/27/2004 20:11	06/27/2004 23:22	03:11	.22	.24	.22	.41	200	03 07:22	
	07/03/2004 17:08	07/04/2004 03:12	10:04	1.59	.84	.60	8.30	1,443	05 17:46	
	07/07/2004 00:04	07/07/2004 01:11	01:07	.58	1.08	.96	5.08	526	02 20:52	
	07/11/2004 20:33	07/11/2004 22:13	01:40	1.08	1.84	1.6	17.1	980	04 19:22	
	07/13/2004 16:15	07/13/2004 18:35	02:20	.19	.52	.26	.46	172	01 18:02	
	07/21/2004 09:40	07/21/2004 14:20	04:40	.23	.76	.44	.91	209	07 15:05	
40	08/02/2004 12:00	08/02/2004 12:26	00:26	.17	.56	—	—	154	11 21:40	
41	08/03/2004 20:10	08/03/2004 23:53	03:43	1.75	3.20	2.14	36.0	1,588	01 07:44	
	08/09/2004 04:51	08/09/2004 09:30	04:39	.33	.52	.40	1.14	299	05 04:58	
42	08/24/2004 20:29	08/25/2004 00:01	03:32	.85	1.76	.92	7.39	771	15 10:59	Rainfall from GMIA
43	08/27/2004 01:30	08/27/2004 02:53	01:23	.37	.64	.54	1.79	336	02 01:29	
44	08/28/2004 01:42	08/28/2004 19:54	18:12	.56	.40	.26	.59	508	00 22:49	
45	09/15/2004 16:03	09/15/2004 22:06	06:03	.28	.40	.26	.63	254	17 20:09	
	10/01/2004 17:04	10/01/2004 23:51	06:47	.22	.24	.20	.37	200	15 18:58	
	10/08/2004 02:44	10/08/2004 13:02	10:18	.14	.08	.06	.07	127	06 02:53	

Table 2–2. Field-blank data summary, hydrodynamic-settling device.

[mg/L, milligrams per liter; µg/L, micrograms per liter; LOD, limit of detection; LOQ, limit of quantification; <, less than]

Constituent	Unit	Blank 1 06/30/03		Blank 2 05/03/04		LOD	LOQ
		Inlet	Outlet	Inlet	Outlet		
Dissolved solids, total	mg/L	<50	<50	<50	<50	50	167
Suspended solids, total recoverable	mg/L	<2	<2	<2	<2	2	7
Suspended sediment, total	mg/L	<2	<2	<2	<2	2	7
Chemical oxygen demand, total	mg/L	<9	<9	<9	55	9	28
Phosphorus, total recoverable	mg/L	<.005	<.005	<.005	<.005	.005	.016
Phosphorus, dissolved	mg/L	<.005	<.005	<.005	<.005	.002	.005
Copper, total recoverable	µg/L	<1	<1	2	1	1	3
Copper, dissolved	µg/L	1.7	1.7	1.6	<1	1	3
Zinc, total recoverable	µg/L	<16	<16	<16	<16	16	50
Zinc, dissolved	µg/L	<16	<16	<16	<16	16	50
Chloride, dissolved	mg/L	.6	1.1	<.6	<.6	.6	2
Calcium, total recoverable	mg/L	<.2	<.2	<.2	<.2	.2	.7
Magnesium, total recoverable	mg/L	<.2	<.2	<.2	<.2	.2	.7

Table 2–3. Hydrodynamic-settling device field-replicate and sample-relative-percent difference data summary

[Rep, replicate; RPD, relative percent difference; %, percent; mg/L, milligrams per liter; µg/L, micrograms per liter; —, no sample processed for event; na, not available]

Parameter	Unit	Site	Event 9			Event 18			Event 42			Objective (%)
			Rep 1a	Rep 1b	RPD (%)	Rep 2a	Rep 2b	RPD (%)	Rep 1a	Rep 1b	RPD (%)	
Dissolved solids, total	mg/L	Inlet	116	116	0	282	286	1	54	60	11	30
		Outlet	178	178	0	394	392	1	128	152	17	
Suspended solids, total recoverable	mg/L	Inlet	186	186	0	312	na	—	70	78	11	30
		Outlet	101	104	3	94	118	23	73	69	6	
Suspended sediment, total	mg/L	Inlet	261	290	11	501	550	9	968	815	17	na
		Outlet	105	102	3	98	100	2	75	79	5	
Chemical oxygen demand, total	mg/L	Inlet	129	133	3	313	362	15	78	53	38	na
		Outlet	119	113	5	223	237	6	84	84	0	
Phosphorus, dissolved	mg/L	Inlet	.10	.10	3	.24	.24	1	.04	.04	0	30
		Outlet	.03	.03	0	.15	.15	0	.03	.03	0	
Phosphorus, total recoverable	mg/L	Inlet	.34	.35	3	.73	.68	7	.20	.23	11	30
		Outlet	.27	.27	1	.49	.48	1	.14	.14	2	
Copper, dissolved	µg/L	Inlet	32.1	32.6	2	75	72.7	3	12.7	13.1	3	25
		Outlet	33.4	32.4	3	34.5	34.9	1	9.9	10	1	
Copper, total recoverable	µg/L	Inlet	113	102	10	202	280	32	111	198	56	25
		Outlet	76	75	1	155	123	23	35	60	53	
Zinc, dissolved	µg/L	Inlet	113	115	2	335	348	4	51	50	2	25
		Outlet	105	110	5	315	325	3	49	52	6	
Zinc, total recoverable	µg/L	Inlet	364	365	0	962	918	5	347	271	25	25
		Outlet	237	247	4	519	523	1	145	172	17	
Chloride, dissolved	mg/L	Inlet	21	21.1	0	78.4	80	2	7	7	0	25
		Outlet	37.3	37.6	1	122	122	0	31.4	31.5	0	
Calcium, total recoverable	mg/L	Inlet	38	47.8	23	48.6	45.3	7	66.4	48	32	25
		Outlet	30.1	31	3	32.3	32.5	1	16.6	16.8	1	
Magnesium, total recoverable	mg/L	Inlet	14.8	20.4	32	20.1	19.3	4	32.2	23.4	32	25
		Outlet	7.4	7.7	4	8.3	8.4	1	4.9	5	2	

Table 2–4. Hydrodynamic-settling-device inlet event start and end time, event volume, percent runoff, and peak discharge.

[mm, month; dd, day; yyyy, year; hh:mm, hour and minutes; in., inch; ft³, cubic foot; ft³/s, cubic foot per second]

Sampled event number	Start date and time (mm/dd/yyyy hh:mm)	End date and time (mm/dd/yyyy hh:mm)	Total rainfall (in.)	Inlet volume (ft³)	Percent runoff	Peak discharge (ft³/s)
1	04/30/2003 13:38	04/30/2003 14:48	0.35	251	79	0.59
2	04/30/2003 22:16	05/01/2003 02:01	1.09	847	86	.46
3	05/04/2003 21:26	05/05/2003 01:51	.72	795	122	.09
4	05/05/2003 04:17	05/05/2003 07:25	.17	130	84	.06
5	05/09/2003 00:27	05/09/2003 04:57	.87	717	91	.13
6	05/20/2003 00:41	05/20/2003 03:14	.19	441	256	.09
7	05/30/2003 18:55	05/30/2003 23:42	.54	665	136	.23
8	06/08/2003 03:26	06/08/2003 16:18	.62	847	150	.60
9	06/27/2003 17:30	06/28/2003 11:15	.57	518	100	.15
10	07/04/2003 07:25	07/06/2003 09:47	.53	492	102	.33
11	07/06/2003 15:08	07/06/2003 16:21	.14	86	68	.06
12	07/08/2003 09:49	07/08/2003 13:45	.33	423	141	.09
13	07/09/2003 23:16	07/09/2003 23:42	.07	43	68	.07
14	07/15/2003 02:59	07/15/2003 05:00	.17	337	218	.08
15	07/30/2003 15:27	07/30/2003 23:37	.19	112	65	.12
16	08/01/2003 02:46	08/03/2003 13:58	.73	484	73	.33
17	08/25/2003 18:44	08/25/2003 19:10	.30	302	111	1.32
18	09/12/2003 15:37	09/12/2003 19:41	.30	156	57	.03
19	09/14/2003 05:30	09/14/2003 12:22	.47	588	138	2.08
20	09/26/2003 16:28	09/26/2003 20:13	.15	112	83	.04
21	10/03/2003 11:19	10/03/2003 12:49	.14	25.9	20	.02
22	10/11/2003 21:53	10/11/2003 23:17	.11	121	121	.05
23	10/14/2003 01:06	10/14/2003 03:19	.27	268	109	.06
24	10/14/2003 08:44	10/14/2003 10:22	.23	138	66	.05
25	10/24/2003 16:46	10/24/2003 22:49	.71	613	95	.16
26	12/28/2003 01:16	12/28/2003 05:49	.22	268	134	.05
27	03/25/2004 23:03	03/26/2004 03:58	.85	311	40	.03
28	03/28/2004 15:24	03/28/2004 20:15	.87	216	27	.03
29	04/17/2004 03:26	04/17/2004 04:25	.24	69	32	.03
30	04/20/2004 16:39	04/21/2004 02:27	1.41	1,028	80	.61
31	05/12/2004 18:27	05/13/2004 03:34	.55	311	62	.11
32	05/20/2004 16:35	05/20/2004 17:41	.24	259	119	1.29
33	05/21/2004 09:04	05/21/2004 10:11	.70	1,020	160	1.81
34	05/30/2004 11:00	05/31/2004 03:45	.68	259	42	.35
35	06/10/2004 11:16	06/11/2004 12:08	1.72	717	46	.07
36	06/14/2004 11:29	06/14/2004 12:16	.82	1,028	138	2.64
37	06/16/2004 19:47	06/16/2004 20:14	.10	78	86	.02
38	06/24/2004 11:32	06/24/2004 12:18	.23	35	17	.02
39	06/27/2004 21:51	06/27/2004 23:25	.22	69	35	.02
40	08/02/2004 12:03	08/02/2004 12:29	.17	354	230	1.01
41	08/03/2004 20:16	08/04/2004 00:06	1.75	2,514	158	2.44
42	08/24/2004 20:32	08/25/2004 00:09	.85	449	58	1.06

Table 2–4. Hydrodynamic-settling-device inlet event start and end time, event volume, percent runoff, and peak discharge.—Continued

[mm, month; dd, day; yyyy, year; hh:mm, hour and minutes; in., inch; ft³, cubic foot; ft³/s, cubic foot per second]

Sampled event number	Start date and time (mm/dd/yyyy hh:mm)	End date and time (mm/dd/yyyy hh:mm)	Total rainfall (in.)	Inlet volume (ft³)	Percent runoff	Peak discharge (ft³/s)
43	08/27/2004 01:39	08/27/2004 03:05	.37	147	44	.90
44	08/28/2004 01:47	08/28/2004 20:13	.56	285	56	.05
45	09/15/2004 16:04	09/15/2004 21:49	.28	78	31	.05
Average			.51	422	94	.44

Table 2–5.   Event mean solids and sediment concentrations during testing of the hydrodynamic-settling device.

[All concentrations in milligrams per liter; —, no sample processed for event; <, less than]

Sampled event number	Dissolved solids, total		Suspended solids, total recoverable		Suspended sediment, total	
	Inlet	Outlet	Inlet	Outlet	Inlet	Outlet
1	224	408	494	250	559	256
2	54	84	79	87	91	86
3	100	118	106	38	161	36
4	68	92	39	13	40	14
5	80	60	89	28	98	27
6	86	186	130	27	—	—
7	88	182	114	70	118	69
8	<50	—	47	64	50.3	64
9	116	178	186	104	290	102
10	80	128	59	43	72	39
11	118	78	70	10	—	—
12	118	128	117	54	123	55
13	108	112	73	15	—	—
14	168	168	63	15	68	14
15	468	200	185	17	223	15
16	64	166	93	90	110	84
17	96	260	66	216	76	212
18	286	394	—	—	550	98
19	<50	—	55	150	79	153
20	202	202	330	8	300	10
21	188	354	108	20	136	21
22	784	746	162	14	142	15
23	82	242	46	33	57	26
24	106	126	128	39	135	34
25	56	128	98	83	106	83
26	2,910	14,500	192	110	194	91
27	184	840	163	139	177	140
28	162	526	272	92	287	90
29	120	1430	113	107	127	106
30	94	440	115	123	123	128
31	80	146	70	42	79	41
32	82	120	97	93	125	92
33	<50	<50	29	35	30	35
34	60	62	44	26	57	27
35	—	<50	84	38	82	36
36	<50	<50	33	95	40	109
37	90	92	73	22	92	22
38	76	172	88	41	90	40
39	60	102	77	51	78	50
40	138	166	432	178	624	178
41	<50	<50	74	87	216	87

Table 2–5. Event mean solids and sediment concentrations during testing of the hydrodynamic-settling device.—Continued

[All concentrations in milligrams per liter; —, no sample processed for event; <, less than]

Sampled event number	Dissolved solids, total		Suspended solids, total recoverable		Suspended sediment, total	
	Inlet	Outlet	Inlet	Outlet	Inlet	Outlet
42	60	152	78	69	815	79
43	<50	—	48	39	85	41
44	66	60	33	12	82	82
45	178	266	104	79	135	82
Count	44	42	44	44	42	42
Average	213	627	117	67	170	73
Median	98	167	89	47	114	67
Geometric mean	122	205	93	48	124	55
Standard deviation	468	2,326	98	55	170	55
Coefficient of variation	2.20	3.71	.83	.82	1.00	.75
Maximum	2,910	14,500	494	250	815	256
Minimum	<50	<50	29	8	30	10

Table 2–6. Event mean concentrations for specific properties and constituents during testing of the hydrodynamic-settling device.

[mg/L, milligrams per liter; µg/L, micrograms per liter; —, not analyzed for event; PAH, polycyclic aromatic hydrocarbon]

Sampled event number	Phosphorus, dissolved (mg/L)		Phosphorus, total recoverable (mg/L)		Chemical oxygen demand (mg/L)		Copper, dissolved (µg/L)		Copper, total recoverable (µg/L)		Zinc, dissolved (µg/L)		Zinc, total recoverable (µg/L)		Chloride, dissolved (mg/L)		Calcium, total recoverable (mg/L)		Magnesium, total recoverable (mg/L)		Total hardness (mg/L)		¹PAH (µg/L)	
	Inlet	Outlet	Inlet	Outlet	Inlet	Outlet	Inlet	Outlet	Inlet	Outlet	Inlet	Outlet	Inlet	Outlet	Inlet	Outlet	Inlet	Outlet	Inlet	Outlet	Inlet	Outlet	Inlet	Outlet
1	0.01	0.02	0.06	0.08	27	39	5	9	25	34	46	68	121	157	10	23	20	18	9	7	86	72	9.3	9.2
3	.109	.056	.226	.115	69	69	14	11	64	24	100	78	268	131	32	40	25	14	9	3	99	47	14.6	4.7
5	.016	.007	.086	.041	58	25	8	5	29	13	64	48	158	84	20	17	20	9	7	2	78	33	—	—
7	.029	.023	.171	.120	66	1350	16	19	56	52	78	96	222	171	18	55	18	23	8	5	76	78	9.5	6.6
8	.038	.028	.098	.128	36	60	10	13	26	33	43	62	102	132	11	33	10	21	4	5	39	75	—	—
9	.098	.033	.354	.269	129	119	33	32	102	75	115	110	365	247	21	38	48	31	20	8	203	106	18.7	5.4
18	.241	.154	.679	.484	313	223	73	35	280	123	348	325	918	523	80	122	45	32	19	8	193	115	—	—
19	.032	.024	.113	.203	31	85	9	10	35	64	42	70	151	266	9	35	10	20	5	9	45	89	—	—
23	.072	.030	.128	.152	52	86	26	33	27	33	17	81	117	130	20	81	15	19	4	3	53	48	—	—
24	.053	.048	.184	.116	90	55	75	32	77	31	48	33	243	101	35	37	20	15	8	3	82	60	8.6	12.3
25	.042	.111	.140	.174	53	81	12	42	51	51	47	165	182	192	12	37	17	17	7	6	71	65	26.8	27.0
27	.021	.017	.134	.136	76	82	12	16	64	73	35	61	244	280	78	481	29	32	11	10	118	121	51.1	15.5
28	.017	.009	.182	.089	101	55	14	12	99	43	85	59	406	193	56	276	47	24	21	7	202	87	—	—
29	.056	.019	.161	.165	72	137	25	43	21	36	108	156	240	306	40	792	20	43	7	8	80	141	—	—
31	.058	.028	.131	.121	60	59	13	19	41	39	53	70	188	147	22	51	13	12	5	3	53	43	—	—
32	.037	.012	.153	.140	57	56	16	14	55	46	60	67	247	173	18	38	16	16	7	6	67	63	—	—
41	.024	.071	.098	.109	33	51	8	8	36	28	33	38	126	126	3	6	19	12	10	5	86	52	13.2	15.1
42	.042	.030	.225	.139	78	84	13	10	198	60	50	52	271	172	7	32	48	17	23	5	216	62	20.7	5.6
Count	18	18	18	18	18	18	18	18	18	18	18	18	18	18	18	18	18	18	18	18	18	18	9	9
Average	.056	.040	.185	.154	78	151	21	20	71	48	76	91	254	196	27	122	24	21	10	6	103	75	19.1	11.3
Median	.040	.028	.147	.132	63	75	14	15	53	41	52	69	231	172	20	38	20	18	8	5	81	68	14.6	9.2
Geometric mean	.042	.029	.156	.136	65	84	16	17	54	42	60	77	217	178	20	55	21	19	9	5	89	70	16.2	9.6
Standard deviation	.053	.038	.140	.096	64	303	20	12	67	25	73	68	185	102	23	204	13	9	6	2	59	30	13.4	7.2
Coefficient of variation	.963	.953	.758	.621	0.8	2.0	1.0	0.6	0.9	0.5	1.0	0.8	0.7	0.5	0.8	1.7	0.5	0.4	0.6	0.4	0.6	0.4	0.7	.64
Maximum	.241	.154	.679	.484	313	1,350	75	43	280	123	348	325	918	523	80	792	48	43	23	10	216	141	51.1	27.0
Minimum	.014	.007	.062	.041	27	25	5	5	21	13	17	33	102	84	3	6	10	9	4	2	39	33	8.6	4.7

¹Summary statistics were only computed for total PAH not of the individual constituents.

Table 2–7. Event-mean polycyclic aromatic hydrocarbon concentrations during testing of the hydrodynamic-settling device.

[All concentrations in micrograms per liter; —, no data; <, less than]

Sampled event number	1-Methylnaphthalene	2-Methylnaphthalene	Fluorene	Acenaphthene	Acenaphthylene	Anthracene	Benzo[a]anthracene	Benzo[a]pyrene	Benzo[b]fluoranthene	Benzo[g,h,i]perylene	Benzo[k]fluoranthene	Chrysene	Dibenzo[a,h]anthracene	Fluoranthene	Indeno[1,2,3-cd]pyrene	Phenanthrene	Pyrene	Naphthalene
Inlet																		
2	<0.046	<0.034	<0.2	<0.06	<0.07	0.058	0.350	0.52	0.86	0.64	0.39	0.73	1.30	1.70	0.61	0.70	1.20	<0.038
3	<.046	.480	<.2	<.06	<.07	.140	.640	.93	1.50	1.20	.67	1.30	.20	3.00	1.10	1.50	2.10	.075
7	<.046	<.034	<.2	<.06	<.07	.055	.370	.63	1.10	.85	.47	.89	.13	1.90	.80	.64	1.30	<.038
9	.047	.060	<.2	<.06	<.07	.134	.779	1.30	2.10	1.60	.93	1.80	.25	3.80	1.50	1.42	2.70	.084
25	<.046	.042	—	.092	—	.170	.600	—	—	.91	.55	.98	—	2.40	.82	—	1.90	.068
27	<.046	.052	<.2	.088	<.07	.320	1.100	1.60	2.70	2.40	1.20	2.30	.40	5.80	2.30	2.10	4.20	.085
28	.076	.096	<.2	.170	<.07	.560	2.200	3.30	5.00	4.40	2.30	4.20	.70	11.0	4.20	4.50	8.10	.170
41	<.046	<.049	<.2	<.06	<.11	.210	.800	1.00	1.20	1.00	.62	1.00	<.20	2.60	0.93	1.10	2.20	<.042
42	<.046	<.049	<.5	<.06	<.11	.410	1.300	1.60	1.80	1.50	.88	1.60	<.30	4.30	1.40	2.10	3.50	.059
Outlet																		
2	<.046	<.034	<.2	<.06	<.07	.062	.340	.54	1.10	.77	.47	.87	.14	1.90	.75	.70	1.30	.05
3	<.046	<.034	<.2	<.06	<.07	<.021	.160	.25	.53	.39	.22	.44	.10	1.00	.37	.34	.68	<.038
7	<.046	<.034	<.2	<.06	<.07	.036	.260	.42	.77	.59	.33	.62	.10	1.40	.53	.43	.96	<.038
9	<.046	<.034	<.2	<.06	<.07	<.021	—	.36	.93	.68	.38	.75	.10	1.30	.63	—	—	<.038
25	.050	.058	—	—	—	.220	.840	—	—	1.40	.83	1.40	—	3.40	1.30	—	2.70	.09
27	<.046	.078	<.2	.073	<.07	.220	.990	1.50	3.00	2.60	1.30	2.40	.40	5.90	2.50	1.60	4.20	.11
28	<.046	<.034	<.2	<.06	<.07	.150	.640	.98	1.60	1.50	.74	1.30	.25	3.30	1.40	1.00	2.40	.05
41	<.064	<.049	<.5	<.06	<.11	.300	1.000	1.20	1.30	1.20	.70	1.10	<.26	2.90	1.10	1.40	2.40	<.042
42	<.064	<.049	<.5	<.06	<.11	.071	.300	.43	.60	.54	.28	.45	<.1	.96	.49	.35	.78	<.042

Table 2–8.   Sum of loads for suspended solids and suspended sediment during testing of the hydrodynamic-settling device.

[All data in pounds; —, no sample processed for event; SOL, sum of loads]

Sampled event number	Dissolved solids, total		Suspended solids, total		Suspended sediment, total	
	Inlet	Outlet	Inlet	Outlet	Inlet	Outlet
1	3.5	6.4	7.8	3.9	8.8	4.0
2	2.9	4.5	4.2	4.6	4.8	4.6
3	5.0	5.9	5.3	1.9	8.0	1.8
4	.6	.7	.3	.1	.3	.1
5	3.6	2.7	4.0	1.3	4.4	1.2
6	2.4	5.2	3.6	.7	—	—
7	3.7	7.6	4.8	2.9	4.9	2.9
8	—	—	2.5	3.4	2.7	3.4
9	3.8	5.8	6.1	3.4	9.4	3.3
10	2.5	4.0	1.8	1.3	2.2	1.2
11	.6	.4	.4	.1	—	—
12	3.1	3.4	3.1	1.4	3.3	1.5
13	.3	.3	.2	.0	—	—
14	3.6	3.6	1.3	.3	1.4	.3
15	3.3	1.4	1.3	.1	1.6	.1
16	1.9	5.0	2.8	2.7	3.3	2.6
17	1.8	4.9	1.3	4.1	1.4	4.0
18	2.8	3.9	—	—	5.4	1.0
19	—	—	2.0	5.5	2.9	5.7
20	1.8	1.8	2.9	.1	2.6	.1
21	1.4	2.7	.8	.2	1.0	.2
22	13.2	12.6	2.7	.2	2.4	.3
23	.7	2.1	.4	.3	.5	.2
24	4.1	4.9	4.9	1.5	5.2	1.3
25	3.4	7.7	5.9	5.0	6.4	5.0
26	49.0	244.1	3.2	1.9	3.3	1.5
27	3.6	16.4	3.2	2.7	3.5	2.7
28	2.2	7.1	3.7	1.2	3.9	1.2
29	.5	6.2	.5	.5	.6	.5
30	6.1	28.4	7.4	7.9	7.9	8.3
31	1.6	2.9	1.4	.8	1.5	.8
32	1.3	2.0	1.6	1.5	2.0	1.5
33	—	—	2.8	3.4	2.9	3.4
34	.7	.7	.5	.3	.6	.3
35	—	—	3.8	1.7	3.7	1.6
36	—	—	2.1	6.1	2.6	7.0
37	.4	.4	.4	.1	.4	.1
38	.2	.4	.2	.1	.2	.1
39	.3	.4	.3	.2	.3	.2
40	3.1	3.7	9.6	4.0	13.9	4.0
41	—	—	11.7	13.7	34.1	13.7

Table 2–8.   Sum of loads for suspended solids and suspended sediment during testing of the hydrodynamic-settling device.—Continued

[All data in pounds; —, no sample processed for event; SOL, sum of loads]

Sampled event number	Dissolved solids, total		Suspended solids, total		Suspended sediment, total	
	Inlet	Outlet	Inlet	Outlet	Inlet	Outlet
42	1.7	4.3	2.2	1.9	23.0	2.2
43	—	—	.4	.4	.8	.4
44	1.2	1.1	.6	.2	1.47	1.47
45	.9	1.3	.5	.4	.7	.4
Total load	143	417	127	94	182	92
SOL	−192		25		49	

Table 2–9. Sum of loads for other constituents during testing of the hydrodynamic-settling device.

[All loads in pounds; —, no sample processed for event; SOL, sum of loads; PAH, polycyclic aromatic hydrocarbon]

Sampled event number	Phosphorus, dissolved		Phosphorus, total recoverable		Chemical oxygen demand		Copper, dissolved		Copper, total recoverable		Zinc, dissolved		Zinc, total recoverable		Chloride, dissolved		[1]PAH, total (1/2 detection)	
	Inlet	Outlet	Inlet	Outlet	Inlet	Outlet	Inlet	Outlet	Inlet	Outlet	Inlet	Outlet	Inlet	Outlet	Inlet	Outlet	Inlet	Outlet
1	0.001	0.001	0.003	0.004	1.4	2.1	0.0003	0.0005	0.0013	0.0018	0.0024	0.0036	0.0064	0.0083	0.0005	0.0012	0.0005	0.0005
3	.0054	.0028	.0112	.0057	3.4	3.4	.0007	.0006	.0032	.0012	.0050	.0039	.0133	.0065	.0016	.0020	.0007	.0002
5	.0007	.0003	.0038	.0018	2.6	1.1	.0004	.0002	.0013	.0006	.0029	.0021	.0071	.0038	.0009	.0008	—	—
7	.0012	.0010	.0071	.0050	2.7	56.1	.0007	.0008	.0023	.0022	.0032	.0040	.0092	.0071	.0007	.0023	.0004	.0003
8	.0020	.0015	.0052	.0068	1.9	3.2	.0005	.0007	.0014	.0017	.0023	.0033	.0054	.0070	.0006	.0017	—	—
9	.0032	.0011	.0115	.0087	4.2	3.9	.0011	.0010	.0033	.0024	.0037	.0036	.0118	.0080	.0007	.0012	.0006	.0002
18	.0023	.0015	.0066	.0047	3.0	2.2	.0007	.0003	.0027	.0012	.0034	.0032	.0089	.0051	.0008	.0012	—	—
19	.0012	.0009	.0041	.0074	1.1	3.1	.0003	.0004	.0013	.0023	.0015	.0026	.0055	.0098	.0003	.0013	—	—
23	.0012	.0005	.0021	.0025	.9	1.4	.0004	.0005	.0005	.0006	.0003	.0014	.0020	.0022	.0003	.0013	—	—
24	.0005	.0004	.0016	.0010	.8	.5	.0006	.0003	.0007	.0003	.0004	.0003	.0021	.0009	.0003	.0003	.0003	.0005
25	.0016	.0042	.0054	.0067	2.0	3.1	.0005	.0016	.0020	.0020	.0018	.0063	.0070	.0074	.0004	.0014	.0005	.0005
27	.0004	.0003	.0026	.0026	1.5	1.6	.0002	.0003	.0012	.0014	.0007	.0012	.0047	.0054	.0015	.0093	.0007	.0002
28	.0002	.0001	.0025	.0012	1.4	.7	.0002	.0002	.0013	.0006	.0011	.0008	.0055	.0026	.0008	.0037	—	—
29	.0002	.0001	.0007	.0007	.3	.6	.0001	.0002	.0001	.0002	.0005	.0007	.0010	.0013	.0002	.0034	—	—
31	.0011	.0005	.0025	.0023	1.2	1.1	.0002	.0004	.0008	.0008	.0010	.0014	.0036	.0029	.0004	.0010	—	—
32	.0006	.0002	.0025	.0023	.9	.9	.0003	.0002	.0009	.0007	.0010	.0011	.0040	.0028	.0003	.0006	—	—
41	.0038	.0111	.0154	.0171	5.2	8.0	.0012	.0012	.0056	.0044	.0052	.0060	.0198	.0198	.0005	.0010	.0021	.0024
42	.0012	.0008	.0063	.0039	2.2	2.4	.0004	.0003	.0056	.0017	.0014	.0015	.0076	.0048	.0002	.0009	.0006	.0002
Total	.0276	.0284	.0943	.0847	36.7	95.3	.0087	.0097	.0354	.0259	.0378	.0466	.1249	.1055	.0110	.0347	.0064	.0049
SOL	-3.0	10	-160	-11	27	-23	16	-216	23									

[1]Summing of PAH for calculation of the event mean concentration was done in three ways: (1) using one-half the detection limit for less than detections, (2) using zero for less than detections, and (3) using the limit of detection value. The three summing methods resulted in means that were in ± 5 percent of one-half of the applicable detection limit.

Appendix 3.   Stormwater-Filtration Device

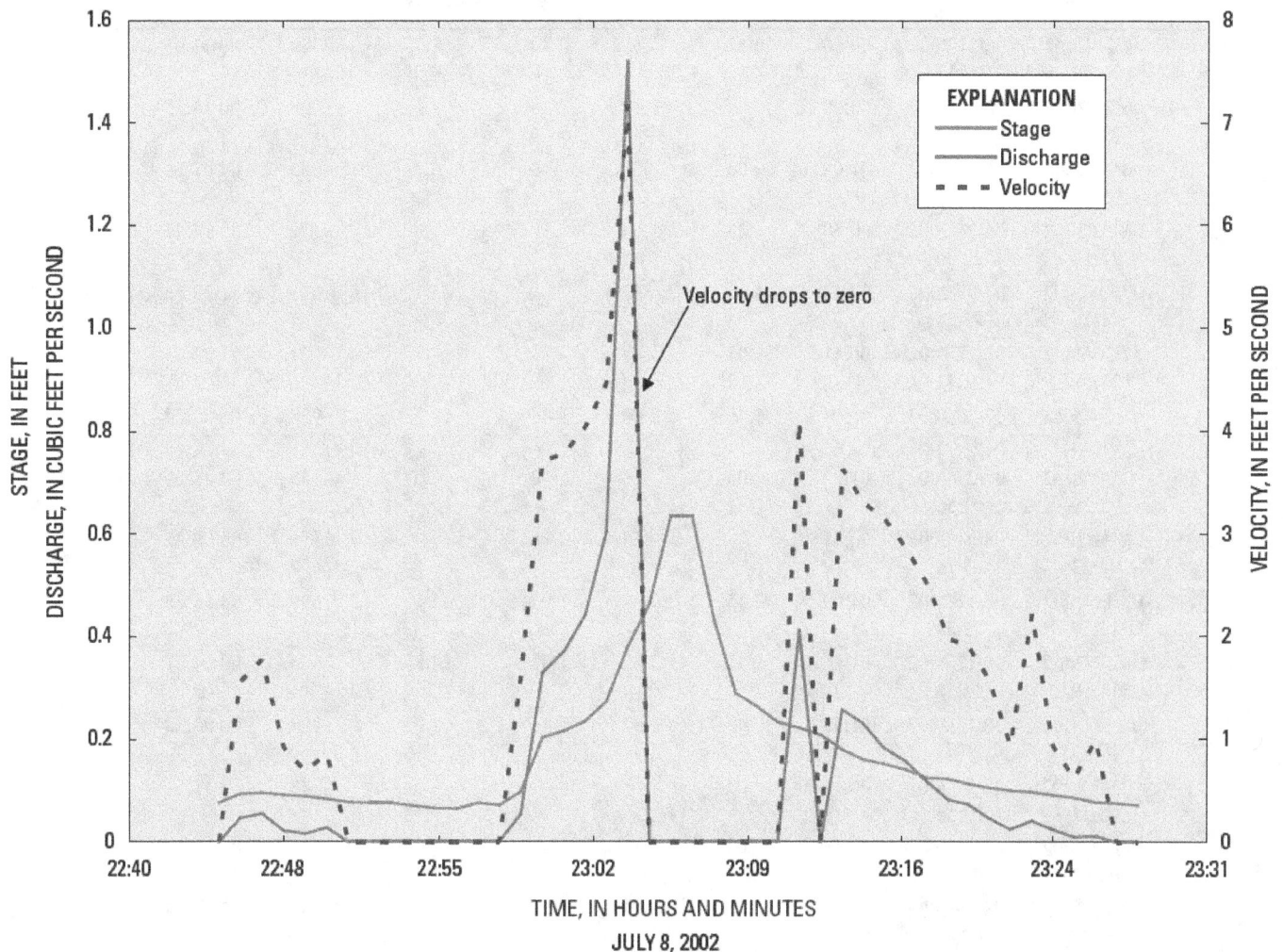

Figure 3–1.   Velocity dropout for the inlet area-velocity flowmeter during a high event flow at the stormwater-filtration device. (The dropout lasted for 8 minutes of the event.)

Table 3–1.   Rainfall data for monitored event, stormwater-filtration device, Milwaukee, Wisconsin.

[mm, month; dd, day; yyyy, year; hh:mm, hour and minutes; in., inch; min, minute; in/h, inch per hour; ft-lb/acre/in/h, foot-pound per acre per inch per hour; ft³, cubic foot; GMIA, General Mitchell International Airport]

Sampled event number	Start date and time (mm/dd/yyyy hh:mm)	End date and time (mm/dd/yyyy hh:mm)	Rainfall duration (hh:mm)	Total rainfall (in.)	Max 15-min intensity (in/h)	Max 30-min intensity (in/h)	Erosivity index (hundreds of ft-lb/acre/in)	Rainfall volume (ft³)	Antecedent dry times (dd hr:mm)	Comments
1	06/21/2002 06:54	06/21/2002 07:14	00:20	0.52	0.56	0.40	1.7	359	06 11:06	
	06/21/2002 21:29	06/21/2002 21:49	00:20	.04	.12	—	—	28	00 14:15	
	06/26/2002 05:15	06/26/2002 06:23	01:08	.08	.20	.10	0.1	55	04 07:26	
2	06/26/2002 21:10	06/26/2002 22:00	00:50	.25	.64	.32	.8	172	00 14:47	
3	07/08/2002 21:16	07/08/2002 23:58	02:42	1.5	.12	—	25.1	1,035	11 23:16	
	07/20/2002 22:11	07/20/2002 22:22	00:11	.03	—	—	—	21	11 22:13	
	07/26/2002 01:26	07/26/2002 03:04	01:38	1.36	2.40	1.44	19.3	938	05 03:04	
	07/29/2002 02:29	07/29/2002 03:56	01:27	.08	.16	.10	.07	55	02 23:25	
4	08/04/2002 04:33	08/04/2002 05:14	00:41	.2	.64	.36	.7	138	06 00:37	
	08/12/2002 19:39	08/12/2002 23:48	04:09	2.45	2.64	1.92	45.6	1,690	08 14:25	
	08/13/2002 14:19	08/13/2002 15:39	01:20	.92	2.52	1.62	14.8	635	00 14:31	
	08/13/2002 19:17	08/13/2002 20:49	01:32	.43	.52	.46	1.7	297	00 03:38	
	08/17/2002 08:41	08/17/2002 08:55	00:14	.07	—	—	—	48	03 11:52	
	08/19/2002 04:07	08/19/2002 04:22	00:15	.03	.12	—	—	21	01 19:12	
	08/19/2002 06:37	08/19/2002 07:47	01:10	.05	.12	.06	.0	34	00 02:15	
5	08/21/2002 20:08	08/22/2002 12:07	15:59	1.67	2.24	1.46	16.7	1,152	02 12:21	
	08/24/2002 03:18	08/24/2002 03:29	00:11	.05	—	—	—	34	01 15:11	
6	09/02/2002 05:24	09/02/2002 08:48	03:24	1.25	1.36	.94	10.6	862	09 01:55	
7	09/02/2002 23:23	09/02/2002 23:42	00:19	.32	1.16	—	—	221	00 14:35	
	09/14/2002 18:45	09/14/2002 18:55	00:10	.04	—	—	—	28	11 19:03	
8	09/18/2002 05:25	09/18/2002 10:19	04:54	.37	.56	.34	1.3	255	03 10:30	
	09/19/2002 14:34	09/19/2002 15:37	01:03	.60	1.24	1.00	5.7	414	01 04:15	

Table 3–1. Rainfall data for monitored event, stormwater-filtration device, Milwaukee, Wisconsin.—Continued

[mm, month; dd, day; yyyy, year; hh:mm, hour and minutes; in., inch; min, minute; in/h, inch per hour; ft-lb/acre/in/h, foot-pound per acre per inch per hour; ft³, cubic foot; GMIA, General Mitchell International Airport]

Sampled event number	Start date and time (mm/dd/yyyy hh:mm)	End date and time (mm/dd/yyyy hh:mm)	Rainfall duration (hh:mm)	Total rainfall (in.)	Max 15-min intensity (in/h)	Max 30-min intensity (in/h)	Erosivity index (hundreds of ft-lb/acre/in)	Rainfall volume (ft³)	Antecedent dry times (dd hr:mm)	Comments
	09/20/2002 09:33	09/20/2002 12:28	02:55	0.10	0.12	0.06	0.1	69	00 17:56	
9	09/29/2002 00:49	09/29/2002 08:43	07:54	.78	.40	.34	2.2	538	08 12:21	
	10/02/2002 00:45	10/02/2002 05:11	04:26	.79	.84	.70	4.8	545	02 16:02	
	10/02/2002 20:12	10/02/2002 23:09	02:57	.09	.16	.10	.1	62	00 15:01	
	10/04/2002 09:35	10/04/2002 13:10	03:35	.54	.80	.48	2.3	372	01 10:26	
	10/12/2002 14:20	10/12/2002 16:39	02:19	.05	.08	.06	.0	34	08 01:10	
	10/18/2002 07:04	10/18/2002 12:54	05:50	.22	.08	.08	.2	152	05 14:25	
	10/24/2002 02:25	10/24/2002 08:47	06:22	.13	.04	.04	.0	90	05 13:31	
	10/25/2002 01:46	10/25/2002 11:20	09:34	.49	.12	.10	.4	338	00 16:59	
	11/05/2002 10:44	11/05/2002 14:51	04:07	.12	.08	.06	.1	83	10 23:24	
	11/11/2002 03:51	11/11/2002 04:59	01:08	.05	.08	.06	.0	34	05 13:00	
	11/11/2002 09:00	11/11/2002 10:50	01:50	.07	.08	.04	.0	48	00 04:01	
	11/18/2002 17:02	11/18/2002 22:08	05:06	.24	.16	.10	.2	166	07 06:12	
10	11/21/2002 05:11	11/21/2002 10:24	05:13	.27	.16	.14	.3	186	02 07:03	
11	12/18/2002 01:18	12/18/2002 06:39	05:21	.37	.36	.30	1.0	255	26 14:54	
	12/18/2002 13:19	12/18/2002 18:38	05:19	.16	.12	.12	.2	110	00 06:40	
12	03/19/2003 12:51	03/19/2003 17:21	04:30	.45	.36	.24	.9	310	30 18:13	
	03/19/2003 20:51	03/19/2003 21:14	00:23	.03	.08	—	—	21	00 03:30	
	03/28/2003 08:44	03/28/2003 15:57	07:13	.32	.36	.28	.8	221	08 11:30	
	03/31/2003 19:06	03/31/2003 20:16	01:10	.06	.08	.06	.0	41	03 03:09	
13	04/04/2003 00:21	04/04/2003 02:37	02:16	.19	.24	.16	.3	131	03 04:05	
13	04/04/2003 07:11	04/04/2003 09:10	01:59	0.18	0.40	0.24	0.4	124	00 04:34	

Table 3–1.   Rainfall data for monitored event, stormwater-filtration device, Milwaukee, Wisconsin.—Continued

[mm, month; dd, day; yyyy, year; hh:mm, hour and minutes; in., inch; min, minute; in/h, inch per hour; ft-lb/acre/in/h, foot-pound per acre per inch per hour; ft³, cubic foot; GMIA, General Mitchell International Airport]

Sampled event number	Start date and time (mm/dd/yyyy hh:mm)	End date and time (mm/dd/yyyy hh:mm)	Rainfall duration (hh:mm)	Total rainfall (in.)	Max 15-min intensity (in/h)	Max 30-min intensity (in/h)	Erosivity index (hundreds of ft-lb/acre/in)	Rainfall volume (ft³)	Antecedent dry times (dd hr:mm)	Comments
	04/04/2003 14:48	04/04/2003 16:06	01:18	.10	.20	.14	.1	69	00 05:38	
	04/06/2003 12:47	04/06/2003 16:19	03:32	.06	.04	.02	.0	41	01 20:41	
	04/08/2003 11:54	04/08/2003 18:02	06:08	.13	.04	.04	.0	90	01 19:35	
	04/09/2003 09:19	04/09/2003 10:54	01:35	.05	.04	.04	.0	34	00 15:17	
14	04/19/2003 05:39	04/19/2003 07:59	02:20	.40	.64	.40	1.4	276	09 18:45	
14	04/19/2003 15:12	04/19/2003 17:03	01:51	.18	.32	.30	.5	124	00 07:13	
	04/20/2003 06:50	04/20/2003 07:10	00:20	.07	.24	—	—	48	00 13:47	
	04/21/2003 08:48	04/21/2003 10:10	01:22	.03	.04	.04	.0	21	01 01:38	
	04/30/2003 07:54	04/30/2003 08:36	00:42	.08	.12	.12	.1	55	08 21:44	
	04/30/2003 13:30	04/30/2003 14:30	01:00	.35	.76	.54	1.7	241	00 04:54	
	04/30/2003 22:08	05/01/2003 01:38	03:30	1.09	1.00	.88	8.4	752	00 07:38	
	05/01/2003 11:19	05/01/2003 14:11	02:52	.08	.12	.08	.1	55	00 09:41	
15	05/04/2003 21:21	05/05/2003 01:26	04:05	.72	.36	.30	1.8	497	03 07:10	
15	05/05/2003 04:14	05/05/2003 09:05	04:51	.17	.24	.20	.3	117	00 02:48	
16	05/07/2003 05:36	05/07/2003 06:35	00:59	.12	.16	.14	.1	83	01 20:31	
16	05/07/2003 11:54	05/07/2003 17:15	05:21	.26	.16	.12	.3	179	00 05:19	
16	05/09/2003 00:12	05/09/2003 04:39	04:27	.87	.60	.42	3.1	600	01 06:57	
	05/11/2003 12:39	05/11/2003 19:57	07:18	.16	.08	.08	.1	110	02 08:00	
	05/14/2003 11:39	05/14/2003 12:53	01:14	.05	.12	.06	.0	34	02 15:42	
	05/14/2003 16:49	05/15/2003 01:14	08:25	.23	.12	.10	.2	159	00 03:56	
	05/15/2003 06:01	05/15/2003 08:03	02:02	.03	.04	.04	.0	21	00 04:47	
	05/20/2003 00:16	05/20/2003 02:41	02:25	0.19	0.16	0.14	0.2	131	04 16:13	

Table 3–1. Rainfall data for monitored event, stormwater-filtration device, Milwaukee, Wisconsin.—Continued

[mm, month; dd, day; yyyy, year; hh:mm, hour and minutes; in., inch; min, minute; in/h, inch per hour; ft-lb/acre/in/h, foot-pound per acre per inch per hour; ft³, cubic foot; GMIA, General Mitchell International Airport]

Sampled event number	Start date and time (mm/dd/yyyy hh:mm)	End date and time (mm/dd/yyyy hh:mm)	Rainfall duration (hh:mm)	Total rainfall (in.)	Max 15-min intensity (in/h)	Max 30-min intensity (in/h)	Erosivity index (hundreds of ft-lb/acre/in)	Rainfall volume (ft³)	Antecedent dry times (dd hr:mm)	Comments
17	05/30/2003 18:54	05/30/2003 23:01	04:07	.54	.52	.32	1.5	372	10 16:13	
	05/31/2003 05:11	05/31/2003 05:28	00:17	.13	.48	—	—	90	00 06:10	
18	06/08/2003 03:26	06/08/2003 14:35	11:09	.62	.80	.54	2.1	428	07 21:58	
19	06/27/2003 17:30	06/27/2003 20:08	02:38	.37	.60	.40	1.3	255	19 02:55	
19	06/28/2003 08:29	06/28/2003 10:55	02:26	.20	.36	.22	.4	138	00 12:21	
20	07/04/2003 07:25	07/04/2003 08:57	01:32	.15	.52	.26	.4	103	05 20:30	
20	07/05/2003 04:33	07/05/2003 06:14	01:41	.31	.36	.32	.8	214	00 19:36	
20	07/06/2003 09:30	07/06/2003 10:08	00:38	.07	.20	.12	.1	48	01 03:16	
	07/06/2003 15:06	07/06/2003 16:19	01:13	.14	.36	.20	.2	97	00 04:58	
	07/06/2003 19:49	07/06/2003 20:02	00:13	.03	—	—	—	21	00 03:30	
	07/07/2003 08:20	07/07/2003 08:49	00:29	.10	.32	—	—	69	00 12:18	
21	07/08/2003 09:49	07/08/2003 13:26	03:37	.33	.24	.20	.6	228	01 01:00	
	07/09/2003 23:14	07/10/2003 00:43	01:29	.07	.24	.12	.1	48	01 09:48	
	07/15/2003 02:56	07/15/2003 04:46	01:50	.17	.20	.12	.2	117	05 02:13	
22	07/21/2003 09:32	07/21/2003 10:14	00:42	.19	.72	.36	.7	131	06 04:46	
23	07/30/2003 15:14	07/30/2003 19:45	04:31	.19	.64	.34	.7	131	09 05:00	Rainfall from GMIA
24	08/01/2003 00:30	08/01/2003 02:54	02:24	.13	.40	.22	.3	90	01 04:45	
24	08/01/2003 06:03	08/01/2003 06:10	00:07	.10	—	—	—	69	00 03:09	
24	08/02/2003 17:38	08/02/2003 17:47	00:09	.09	—	—	—	62	01 11:28	
24	08/03/2003 12:34	08/03/2003 14:21	01:47	.41	.64	.50	1.8	283	00 18:47	
	08/11/2003 22:54	08/11/2003 23:41	00:47	0.11	0.40	0.20	0.2	76	08 08:33	
25	08/25/2003 18:49	08/25/2003 19:36	00:47	.30	1.16	.58	1.8	207	13 19:08	

Table 3–1. Rainfall data for monitored event, stormwater-filtration device, Milwaukee, Wisconsin.—Continued

[mm, month; dd, day; yyyy, year; hh:mm, hour and minutes; in., inch; min, minute; in/h, inch per hour; ft-lb/acre/in/h, foot-pound per acre per inch per hour; ft³, cubic foot; GMIA, General Mitchell International Airport]

Sampled event number	Start date and time (mm/dd/yyyy hh:mm)	End date and time (mm/dd/yyyy hh:mm)	Rainfall duration (hh:mm)	Total rainfall (in.)	Max 15-min intensity (in/h)	Max 30-min intensity (in/h)	Erosivity index (hundreds of ft-lb/acre/in)	Rainfall volume (ft³)	Antecedent dry times (dd hr:mm)	Comments
26	09/12/2003 15:32	09/12/2003 19:21	03:49	.30	.24	.22	.6	207	17 19:56	
27	09/13/2003 07:30	09/13/2003 10:52	03:22	.16	.16	.12	.2	110	00 12:09	
28	09/14/2003 05:22	09/14/2003 11:57	06:35	.47	1.36	.16	.2	324	00 18:30	
29	09/22/2003 02:28	09/22/2003 06:05	03:37	.27	.32	.24	.7	186	07 14:31	
	09/26/2003 16:11	09/26/2003 19:23	03:12	.15	.16	.14	.2	103	04 10:06	
	10/03/2003 10:15	10/03/2003 12:23	02:08	.14	.12	.12	.1	97	06 14:52	
	10/11/2003 21:58	10/12/2003 00:02	02:04	.11	.08	.08	.1	76	08 09:35	
30	10/14/2003 00:17	10/14/2003 03:10	02:53	.27	.20	.16	.4	186	02 00:15	
31	10/14/2003 07:08	10/14/2003 09:49	02:41	.23	.24	.20	.4	159	00 03:58	
32	10/24/2003 16:45	10/24/2003 22:16	05:31	.71	.36	.34	2.0	490	10 06:56	
	11/01/2003 22:06	11/02/2003 08:05	09:59	.63	.32	.24	1.3	435	07 23:50	
33	11/04/2003 16:14	11/04/2003 20:21	04:07	.60	.68	.36	1.4	414	02 08:09	Rainfall from GMIA
	11/17/2003 23:10	11/18/2003 12:11	13:01	1.08	.52	.40	3.6	745	13 02:49	
	11/22/2003 17:26	11/22/2003 21:59	04:33	.12	.12	.08	.1	83	04 05:15	
	11/23/2003 05:37	11/23/2003 15:04	09:27	.13	.12	.10	.1	90	00 07:38	
	12/09/2003 12:30	12/10/2003 16:57	04:27	1.90	.32	.24	3.8	1,310	15 21:26	
	12/16/2003 03:29	12/16/2003 04:58	01:29	.11	.16	.12	.1	76	05 10:32	
	12/28/2003 01:06	12/28/2003 05:34	04:28	.22	.16	.12	.2	152	11 20:08	

Table 3–2. Field-blank data summary, stormwater-filtration device.

[mg/L, milligrams per liter; µg/L, micrograms per liter; LOD, limit of detection; LOQ, limit of quantification; —, no sample processed; <, less than]

Constituent	Unit	Blank 1 4/2/2002		Blank 2 11/11/2002		Blank 3 6/30/2003		LOD	LOQ
		Inlet	Outlet	Inlet	Outlet	Inlet	Outlet		
Suspended solids, total recoverable	mg/L	<2	<2	—	—	<2	<2	2	7
Suspended sediment, total	mg/L	—	—	—	—	<2	<2	2	7
Dissolved solids, total	mg/L	<50	<50	<50	<50	<50	<50	50	167
Total chemical oxygen, demand	mg/L	<9	<9	<9	<9	12	14	9	28
Phosphorus, dissolved	mg/L	—	—	<.005	<.005	<.005	<.005	.005	.016
Phosphorus, total recoverable	mg/L	<.005	<.005	.025	<.005	<.005	<.005	.005	.016
Copper, dissolved	µg/L	<5	<5	<1	<1	1.7	2.3	1	3
Copper, total recoverable	µg/L	<5	<5	<1	<1	2	2	1	3
Zinc, dissolved	µg/L	<16	<16	<16	<16	<16	<16	16	50
Zinc, total recoverable	µg/L	<16	<16	<16	<16	<16	<16	16	50
Chloride, dissolved	mg/L	3.3	<.6	<.6	<.6	.8	<.6	.6	2
Calcium, total recoverable	mg/L	.7	<.2	<.2	<.2	<.2	<.2	.200	.070
Magnesium, total recoverable	mg/L	<.2	<.2	<.2	<.2	<.2	<.2	.200	.070

Table 3–3. Stormwater-filtration device field-replicate and sample relative-percent-difference data summary.

[Rep, replicate; RPD, relative percent difference; %, percent; mg/L, milligrams per liter; µg/L, micrograms per liter; na, not available; —, no sample processed for event; <, less than]

Parameter	Unit	Site	Event 9 Rep 1a	Event 9 Rep 1b	Event 9 RPD (%)	Event 14 Rep 2a	Event 14 Rep 2b	Event 14 RPD (%)	Event 19 Rep 3a	Event 19 Rep 3b	Event 19 RPD (%)	Event 26 Rep 4a	Event 26 Rep 4b	Event 26 RPD (%)	Event 28 Rep 5a	Event 28 Rep 5b	Event 28 RPD (%)	Objective (%)
Dissolved solids, total	mg/L	Inlet	<50	52	na	516	522	-1	90	86	5	212	224	-6	50	<50	na	30
		Outlet	<50	<50	na	722	728	-1	162	160	1	190	194	-2	74	58	24	30
Suspended solids, total recoverable	mg/L	Inlet	—	—	—	778	838	-7	77	96	-22	696	816	-16	35	44	-23	na
		Outlet	—	—	—	378	380	-1	46	47	-2	36	31	15	20	25	-22	30
Suspended sediment, total	mg/L	Inlet	501	681	-30	5,590	4,860	14	368	212	54	3,750	2,410	44	411	306	29	na
		Outlet	39	39	0	373	373	0	47	48	-2	29	32	-10	21	22	-5	na
Chemical oxygen demand. total	mg/L	Inlet	41	45	-9	315	287	9	85	86	-1	298	295	1	51	48	6	30
		Outlet	30	22	31	190	187	2	81	87	-7	162	153	6	50	53	-6	na
Phosphorus, dissolved	mg/L	Inlet	.03	.031	-3	.027	.025	8	.061	.063	-3	.199	.206	-3	.04	.039	3	30
		Outlet	.027	.026	4	.017	.016	6	.059	.058	2	.193	.193	0	.046	.046	0	30
Phosphorus, total recoverable	mg/L	Inlet	.159	.109	37	.502	.555	-10	.235	.32	-31	.625	.584	7	.149	.105	35	30
		Outlet	.067	.065	3	.292	.302	-3	.189	.188	1	.298	.285	4	.098	.098	0	30
Copper, dissolved	µg/L	Inlet	8.9	9.5	-7	27.8	27.6	1	20	21.2	-6	57.5	58.5	-2	49.5	166	-108	25
		Outlet	6.8	8.4	-21	27.1	25.7	5	22.6	23.3	-3	41.7	40.7	2	17.6	18.6	-6	25
Copper, total recoverable	µg/L	Inlet	139	35	120	277	372	-29	48	52	-8	331	258	25	46	133	-97	25
		Outlet	17	18	-6	139	140	-1	44	46	-4	69	68	1	15	15	0	25
Zinc, dissolved	µg/L	Inlet	35	31	12	112	119	-6	81	77	5	358	353	1	46	47	-2	25

Table 3–3. Stormwater-filtration device field-replicate and sample relative-percent-difference data summary.—Continued

[Rep, replicate; RPD, relative percent difference; %, percent; mg/L, milligrams per liter; µg/L, micrograms per liter; na, not available; —, no sample processed for event]

Parameter	Unit	Site	Event 9 Rep 1a	Rep 1b	RPD (%)	Event 14 Rep 2a	Rep 2b	RPD (%)	Event 19 Rep 3a	Rep 3b	RPD (%)	Event 26 Rep 4a	Rep 4b	RPD (%)	Event 28 Rep 5a	Rep 5b	RPD (%)	Objective (%)
Zinc, total recoverable	µg/L	Outlet	22	22	0	84	91	-8	96	92	4	158	153	3	42	43	-2	
		Inlet	134	328	-84	1,380	2,200	-46	198	324	-48	1,370	1,700	-21	296	281	5	25
		Outlet	61	63	-3	539	544	-1	158	156	1	215	208	3	66	67	-2	
Chloride, dissolved	mg/L	Inlet	na	na	—	468	477	-2	16.7	17.1	-2	34.4	33.4	3	5.4	5.3	2	25
		Outlet	na	na	—	661	673	-2	34.8	34.7	0	35.2	34.9	1	—	8.7	na	
Calcium, total recoverable	mg/L	Inlet	16	20.1	-23	434	475	-9	29.3	32	-9	233	217	7	59.7	62.2	-4	25
		Outlet	6.1	6.2	-2	68.2	68.4	0	17.2	17.5	-2	16.3	15.6	4	7	7.1	-1	
Magnesium, total recoverable	mg/l	Inlet	7.8	10.1	-26	174	201	-14	11.2	11.5	-3	122	111	9	22.2	27.1	-20	25
		Outlet	2.5	2.5	0	25.8	26	-1	4.2	4.2	0	4.4	4.2	5	1.9	2	-5	

Table 3–4. Stormwater-filtration-device outlet event start and end time, event volume, percent runoff, and peak discharge.

[mm, month; dd, day; yyyy, year; hh:mm, hour and minutes; in., inch; ft^3, cubic foot; ft^3/s, cubic foot per second]

Sampled event number	Start date and time (mm/dd/yyyy hh:mm)	End date and time (mm/dd/yyyy hh:mm)	Total rainfall (in.)	Volume (ft^3)	Percent runoff	Peak discharge (ft^3/s)
1	06/21/2002 06:54	06/21/2002 07:40	0.52	354	99	1.11
2	06/26/2002 21:10	06/26/2002 22:19	.25	138	80	.28
3	07/08/2002 21:16	07/08/2002 23:41	1.5	1,253	121	1.06
4	08/04/2002 04:35	08/04/2002 05:01	.20	69	50	.20
5	08/21/2002 20:12	08/22/2002 12:37	1.67	968	84	1.12
6	09/02/2002 05:24	09/02/2002 09:48	1.25	648	75	.30
7	09/02/2002 23:26	09/02/2002 23:51	.32	242	110	.38
8	09/18/2002 05:25	09/18/2002 10:25	.37	207	81	.25
9	09/29/2002 02:49	09/29/2002 09:27	.78	233	43	.01
10	11/21/2002 05:15	11/21/2002 11:26	.27	112	60	.08
11	12/18/2002 01:18	12/18/2002 06:02	.37	104	41	.08
12	03/19/2003 13:51	03/19/2003 17:07	.45	302	97	.14
13	04/04/2003 01:01	04/04/2003 09:12	.37	181	71	.26
14	04/19/2003 05:39	04/19/2003 15:55	.58	233	58	.25
15	05/04/2003 21:26	05/05/2003 07:25	.89	337	55	.19
16	05/07/2003 05:42	05/09/2003 04:57	1.25	588	68	.28
17	05/30/2003 18:55	05/30/2003 23:42	.54	207	56	.21
18	06/08/2003 03:26	06/08/2003 16:18	.62	354	83	.34
19	06/27/2003 17:30	06/28/2003 11:15	.57	363	92	.27
20	07/04/2003 07:25	07/06/2003 09:47	.53	622	170	.36
21	07/08/2003 09:49	07/08/2003 13:45	.33	250	110	.17
22	07/21/2003 09:37	07/21/2003 10:08	.19	173	132	.39
23	07/30/2003 15:27	07/30/2003 23:37	.19	61	46	.02
24	08/01/2003 02:46	08/03/2003 13:58	.73	605	120	.33
25	08/25/2003 18:44	08/25/2003 19:10	.30	250	121	.53
26	09/12/2003 15:37	09/12/2003 19:41	.30	156	75	.02
27	09/13/2003 07:34	09/13/2003 11:28	.16	78	70	.01
28	09/14/2003 05:30	09/14/2003 12:22	.47	337	104	.52
29	09/22/2003 02:29	09/22/2003 04:54	.27	207	111	.27
30	10/14/2003 01:06	10/14/2003 03:19	.27	130	70	.14
31	10/14/2003 08:44	10/14/2003 10:22	.23	52	33	.02
32	10/24/2003 16:46	10/24/2003 22:49	.71	225	46	.20
33	11/04/2003 16:14	11/04/2003 19:30	.60	596	144	1.12
Average			.55	322	84	.33

Table 3–5. Event-mean solids and sediment concentrations during testing of the stormwater-filtration device.

[All concentrations in milligrams per liter; —, no sample processed for event; <, less than]

Sampled event number	Dissolved solids, total		Suspended solids, total recoverable		Suspended sediment, total	
	Inlet	Outlet	Inlet	Outlet	Inlet	Outlet
1	—	<50	71	83	372	63
2	<50	<50	76	48	697	12
3	<50	<50	51	28	312	20
4	<50	<50	—	—	476	36
5	<50	<50	—	—	65	19
6	39	38	—	—	324	13
7	<50	<50	—	—	154	13
8	<50	<50	—	—	119	43
9	<50	<50	—	—	140	12
10	—	—	—	—	430	103
11	596	4,170	—	—	770	129
12	—	—	—	—	456	401
13	<50	<50	736	31	3,820	318
14	516	722	778	378	5,590	373
15	78	90	73	34	825	34
16	66	64	79	29	984	29
17	66	126	112	70	1,280	68
18	<50	—	60	40	419	40.4
19	90	162	77	46	368	47
20	60	110	29	30	51	32
21	82	108	57	24	74	23
22	68	110	51	103	24	98
23	208	276	60	36	—	—
24	<50	—	22	36	27	34
25	72	124	68	90	256	90
26	212	190	696	36	3,750	29
27	88	168	30	18	36	19
28	<50	—	50	49	405	49
29	50	80	37	31	484	21
30	50	74	35	20	411	21
31	56	78	53	28	130	24
32	<50	—	67	36	416	33
33	<50	<50	55	73	103	97
Count	30	27	24	24	32	32
Average	141	394	143	58	743	73
Median	72	110	60	36	389	34
Geometric mean	96	149	75	43	307	43
Standard deviation	164	986	230	72	1,255	100
Coefficient of variation	1.2	2.50	1.6	1.2	1.7	1.4
Maximum	596	4,170	778	378	5,590	401
Minimum	39	38	22	18	24	12

Table 3–6. Event-mean concentrations for specific properties and constituents during testing of the stormwater-filtration device.

[mg/L, milligrams per liter; µg/L, micrograms per liter; —, no sample processed for event; <, less than; PAH, polycyclic aromatic hydrocarbon]

Sampled event number	Phosphorus, dissolved (mg/L) Inlet	Outlet	Phosphorus, total recoverable (mg/L) Inlet	Outlet	Chemical oxygen demand (mg/L) Inlet	Outlet	Copper, dissolved (µg/L) Inlet	Outlet	Copper, total recoverable (µg/L) Inlet	Outlet	Zinc, dissolved (µg/L) Inlet	Outlet	Zinc, total recoverable (µg/L) Inlet	Outlet	Chloride, dissolved (mg/L) Inlet	Outlet	Calcium, total recoverable (mg/L) Inlet	Outlet	Magnesium, total recoverable (mg/L) Inlet	Outlet	Total hardness (mg/L) Inlet	Outlet	[1]PAH, total (µg/L) Inlet	Outlet
1	0.041	0.039	0.138	0.097	42	37	<5	<5	41	28	60	34	222	139	5.8	5.2	42.4	14.8	21	6	—	—	—	—
3	.041	.037	.105	.078	39	25	10.0	8.8	34	19	59	51	202	76	4.6	4.6	27.8	6.0	14	2	205	16	—	—
5	.014	.013	.045	.037	18	24	6.1	5.4	15	10	27	20	176	39	4.5	3.4	9.7	4.4	4	2	41	17	1.46	0.99
6	.030	.032	.099	.048	29	24	7.7	7.0	29	10	49	43	198	56	3.2	3.3	54.6	4.4	26	1	242	17	7.90	0.94
8	.059	.046	.135	.101	80	78	21.0	14.2	127	30	87	51	677	109	—	—	16.9	9.7	7	3	72.4	37	—	—
9	.021	.021	.100	.030	28	17	5.0	4.5	16	7	26	16	77	28	3.6	4	9.4	4.0	4	101	40	15	6.56	0.86
11	.035	.029	.327	.202	68	129	13.9	20.4	126	78	59	109	393	302	310	2,590	130.0	47.5	56	9	557	153	138.84	33.49
12	.027	.017	.502	.292	315	190	27.8	27.1	277	139	112	84	1,380	539	468	661	434.0	68.2	174	26	1,800	276	23.89	3.44
15	.057	.043	.170	.080	53	38	11.4	8.7	44	20	64	45	230	91	25	31	61.6	10.5	28	3	267	38	11.87	6.89
17	.045	.028	.200	.138	67	61	16.8	15.2	79	42	67	70	243	145	14	32	39.7	16.6	18	5	173	61	—	—
18	.023	.028	.193	.080	41	36	18.0	7.6	36	23	37	32	117	84	9.4	17	37.3	9.6	18	3	167	36	5.26	2.43
19	.061	.059	.235	.189	85	81	20.0	22.6	48	44	81	96	198	158	17	35	29.3	17.2	11	4	119	60	—	—
21	.048	.049	.162	.107	63	53	13.2	15.1	36	29	57	42	226	79	20	22	12.4	8.9	5	2	51	32	—	—
26	.199	.193	.625	.298	298	162	57.5	41.7	331	69	358	158	1,370	215	34	35	233	16.3	122	4	1,080	59	—	—
28	.020	.027	.100	.095	38	34	5.5	6.2	32	21	26	30	184	106	6.1	9.7	40.8	8.8	20	4	184	37	—	—
29	.043	.054	.152	.098	48	72	9.0	10.5	440	18	42	47	650	69	9	16	73.0	8.3	36	3	331	31	—	—
30	.040	.046	.149	.098	51	50	49.5	17.6	46	15	46	42	296	66	5.4	—	59.7	7.0	22	2	240	25	—	—
Count	17	17	17	17	17	17	16	16	17	17	17	17	17	17	16	15	17	17	17	17	16	16	7	7
Average	.047	.045	.202	.122	80	65	18.3	14.5	103	35	74	57	402	135	58.7	231.3	77.2	15.4	34	11	348	57	27.97	7.01
Median	.041	.037	.152	.098	51	50	13.6	12.4	44	23	59	45	226	91	9.2	17.0	40.8	9.6	20	3	195	37	7.90	2.43
Geomean	.039	.036	.165	.101	58	51	14.1	12.0	61	26	58	48	288	103	14.0	21.9	43.7	10.9	19	4	195	39	10.70	2.85
Standard deviation	.042	.040	.152	.079	87	50	15.2	9.8	125	33	77	36	401	124	132.4	673.4	106.9	16.9	46	24	464	67	49.41	11.87
Coefficient of variation	0.9	0.9	0.7	0.7	1.1	0.8	0.8	0.7	1.2	0.9	1.0	0.6	1.0	0.9	2.3	2.9	1.4	1.1	1.3	2.3	1.3	1.2	0.57	0.59
Maximum	0.199	.193	0.625	0.298	315	190	57.5	41.7	440	139	358	158	1,380	539	468.0	2,590.0	434.0	68.2	174	101	1,800	276	138.84	33.49
Minimum	0.014	0.013	0.045	0.030	17	17	5.0	4.5	15	7	26	16	77	28	3.2	3.3	9.4	4.0	4	1	40	15	1.49	0.86

[1]Summary statistics were only computed for total PAH, not for the individual constituents

Table 3–7. Event-mean polycyclic aromatic hydrocarbon[1] concentrations during testing of the stormwater-filtration device.

[All concentrations in micrograms per liter; <, less than]

Sampled event number	1-Methylnaphthalene	2-Methylnaphthalene	Fluorene	Acenaphthene	Acenaphthylene	Anthracene	Benzo[a]anthracene	Benzo[a]pyrene	Benzo[b]fluoranthene	Benzo[g,h,i]perylene	Benzo[k]fluoranthene	Chrysene	Dibenzo[a,h]anthracene	Fluoranthene	Indeno[1,2,3-cd]pyrene	Phenanthrene	Pyrene	Naphthalene
Inlet																		
5	<0.046	<0.034	<0.2	<0.06	<0.072	<0.021	<0.062	<0.07	0.13	0.120	<0.07	0.1	0.04	0.24	0.12	0.13	0.18	<0.038
6	<.046	<.034	<.2	<.06	<.072	.630	.35	.43	.59	.480	.30	.54	1.0	1.30	.46	.59	1.00	<.038
9	<.046	.035	<.2	<.06	<.072	.071	.30	.37	.51	.450	.26	.47	.80	1.10	.41	.66	.89	.046
12	.290	<.700	.9	<.80	<.72	2.000	7.00	9.30	13.00	9.30	6.10	12.00	<1.8	30.00	8.90	16.00	22.00	.32
15	.074	.081	.3	<.06	<.072	.480	1.30	1.60	1.90	1.50	.94	1.90	.25	4.90	1.30	3.30	3.9	.10
17	<.046	<.034	<.2	<.06	<.072	.140	.64	.89	1.20	.920	.55	1.00	.15	2.40	.83	1.1	1.9	<.038
19	<.046	<.034	<.2	<.06	<.072	.044	.25	.38	0.60	.480	.27	.53	.08	1.10	.42	.04	.84	<.038
Outlet																		
5	<.046	<.034	<.2	<.06	<.072	<.021	<.062	<.07	<.11	<.078	<.07	.05	.04	.14	.12	.12	.10	<.038
6	<.046	<.034	<.2	<.06	<.072	<.021	<.062	<.07	<.11	.08	<.07	.05	.04	.13	.12	.04	.09	<.038
9	<.046	<.034	<.2	<.06	<.072	<.021	<.062	<.07	<.11	<.078	<.07	.04	.04	.10	.12	.09	<.07	<.038
12	.05	.066	<.2	<.06	<.072	.220	1.20	1.70	3.90	2.80	1.70	3.30	.70	7.20	2.7	2.70	5.00	.09
15	<.046	<.034	<.2	<.06	<.072	<.021	.13	.21	.38	.310	.17	.30	.04	.62	.28	.31	.46	<.038
17	<.046	<.034	<.2	<.06	<.072	.095	.41	.53	.70	.550	.33	.62	.10	1.30	.50	.59	.94	<.038
19	<.046	<.034	<.2	<.06	<.072	<.021	.09	.15	.28	.230	<.07	.22	.04	.44	.20	.18	.33	<.038

[1]Summary statistics were only computed for total PAH not of the individual constituents.

Table 3–8. Sum of loads for dissolved solids and suspended sediment during testing of the stormwater-filtration device.

[All data in pounds; —, no sample processed for event; SOL, sum of loads]

Sampled event number	Dissolved solids, total		Suspended solids, total		Suspended sediment, total	
	Inlet	Outlet	Inlet	Outlet	Inlet	Outlet
1	—	—	1.6	1.8	8.3	1.4
2	—	—	.7	.4	6.1	.1
3	—	—	4.0	2.2	24.6	1.6
4	—	—	—	—	2.1	.2
5	—	—	—	—	4.0	1.2
6	1.6	1.5	—	—	13.2	.5
7	—	—	—	—	2.3	.2
8	—	—	—	—	1.6	.6
9	—	—	—	—	2.1	.2
10	—	—	—	—	3.0	.7
11	3.9	27.2	—	—	5.0	.8
12	—	—	—	—	8.7	7.6
13	—	—	8.4	.4	43.6	3.6
14	7.6	10.6	11.4	5.5	82.0	5.5
15	1.7	1.9	1.5	.7	17.5	.7
16	2.4	2.4	2.9	1.1	36.3	1.1
17	.9	1.6	1.5	.9	16.7	.9
18	—	—	1.3	.9	9.3	.9
19	2.1	3.7	1.8	1.0	8.4	1.1
20	2.3	4.3	1.1	1.2	2.0	1.3
21	1.3	1.7	.9	.4	1.2	.4
22	.7	1.2	.6	1.1	.3	1.1
23	.8	1.0	.2	.1	—	—
24	—	—	.8	1.4	1.0	1.3
25	1.1	2.0	1.1	1.4	4.0	1.4
26	2.1	1.9	6.8	.4	36.7	.3
27	.4	.8	.1	.1	.2	.1
28	—	—	1.1	1.0	8.6	1.0
29	.7	1.0	.5	.4	6.3	.3
30	.4	.6	.3	.2	3.3	.2
31	.2	.3	.2	.1	.4	.1
32	—	—	.9	.5	5.9	.5
33	—	—	2.1	2.7	3.9	3.6
Total load	30	64	51.8	25.9	368	40
SOL	−112		50		89	

Table 3-9. Sum of loads for other constituents during testing of the stormwater-filtration device.

[All loads in pounds; —, no sample processed for event; SOL, sum of loads ; PAH, polycyclic aromatic hydrocarbon]

Sampled event number	Phosphorus, dissolved		Phosphorus, total recoverable		Chemical oxygen demand		Copper, dissolved		Copper, total recoverable		Zinc, dissolved		Zinc, total recoverable		Chloride, dissolved		[1]PAH, total (1/2 detection)	
	Inlet	Outlet	Inlet	Outlet	Inlet	Outlet	Inlet	Outlet	Inlet	Outlet	Inlet	Outlet	Inlet	Outlet	Inlet	Outlet	Inlet	Outlet
1	0.0009	0.0009	0.003	0.002	0.93	0.82	0.0001	0.0001	0.0009	0.0006	0.0013	0.0008	0.005	0.003	0.13	0.12	—	—
3	.0032	.0029	.008	.006	3.05	1.96	.0008	.0007	.0027	.0015	.0046	.0040	.016	.006	.36	.36	—	—
5	.0008	.0008	.003	.002	1.09	1.45	.0004	.0003	.0009	.0006	.0016	.0012	.011	.002	.27	.20	0.0001	0.0000
6	.0012	.0013	.004	.002	1.17	.97	.0003	.0003	.0012	.0004	.0020	.0017	.008	.002	.13	.13	.0003	.0001
8	.0008	.0006	.002	.001	1.04	1.01	.0003	.0002	.0016	.0004	.0011	.0007	.009	.001	—	—	.0002	.0006
9	.0005	.0005	.002	.001	.60	.37	.0001	.0001	.0003	.0002	.0006	.0003	.002	.001	.08	.09	.0001	.0000
11	.0006	.0005	.006	.004	1.25	2.36	.0003	.0004	.0023	.0014	.0011	.0020	.007	.006	5.68	47.49	—	—
12	.0002	.0001	.003	.002	2.04	1.23	.0002	.0002	.0018	.0009	.0007	.0005	.009	.003	3.03	4.28	.0020	.0005
15	.0008	.0006	.002	.001	.77	.55	.0002	.0001	.0006	.0003	.0009	.0007	.003	.001	.36	.45	.0004	.0001
17	.0008	.0005	.003	.002	1.16	1.05	.0003	.0003	.0014	.0007	.0012	.0012	.004	.003	.24	.55	.0002	.0001
18	.0003	.0004	.002	.001	.53	.47	.0002	.0001	.0005	.0003	.0005	.0004	.002	.001	.12	.22	—	—
19	.0013	.0013	.005	.004	1.88	1.79	.0004	.0005	.0011	.0010	.0018	.0021	.004	.003	.38	.77	.0001	.0000
21	.0011	.0011	.004	.002	1.43	1.20	.0003	.0003	.0008	.0007	.0013	.0010	.005	.002	.45	.50	—	—
26	.0031	.0030	.010	.005	4.66	2.53	.0009	.0007	.0052	.0011	.0056	.0025	.021	.003	.53	.55	—	—
28	.0002	.0003	.001	.001	.37	.33	.0001	.0001	.0003	.0002	.0003	.0003	.002	.001	.06	.09	—	—
29	.0009	.0011	.003	.002	1.01	1.51	.0002	.0002	.0093	.0004	.0009	.0010	.014	.001	.19	.35	—	—
30	.0005	.0006	.002	.001	.66	.65	.0006	.0002	.0006	.0002	.0006	.0005	.004	.001	—	—	—	—
Total	.0173	.0164	.0643	.0401	23.64	20.25	.0057	.0048	.0314	.0108	.0260	.0209	.125	.0416	12.1	56.15	.0034	.0013
SOL	5	38	14	16	66	20	68	-367.53	59									

[1]Summing of PAH for calculation of the event-mean concentration was done in three ways: (1) using one-half the detection limit for less than detections, (2) using zero for less than detections, and (3) using the limit of detection value. The three summing methods resulted in means that were ± 5 percent of one-half of the applicable detection limit.